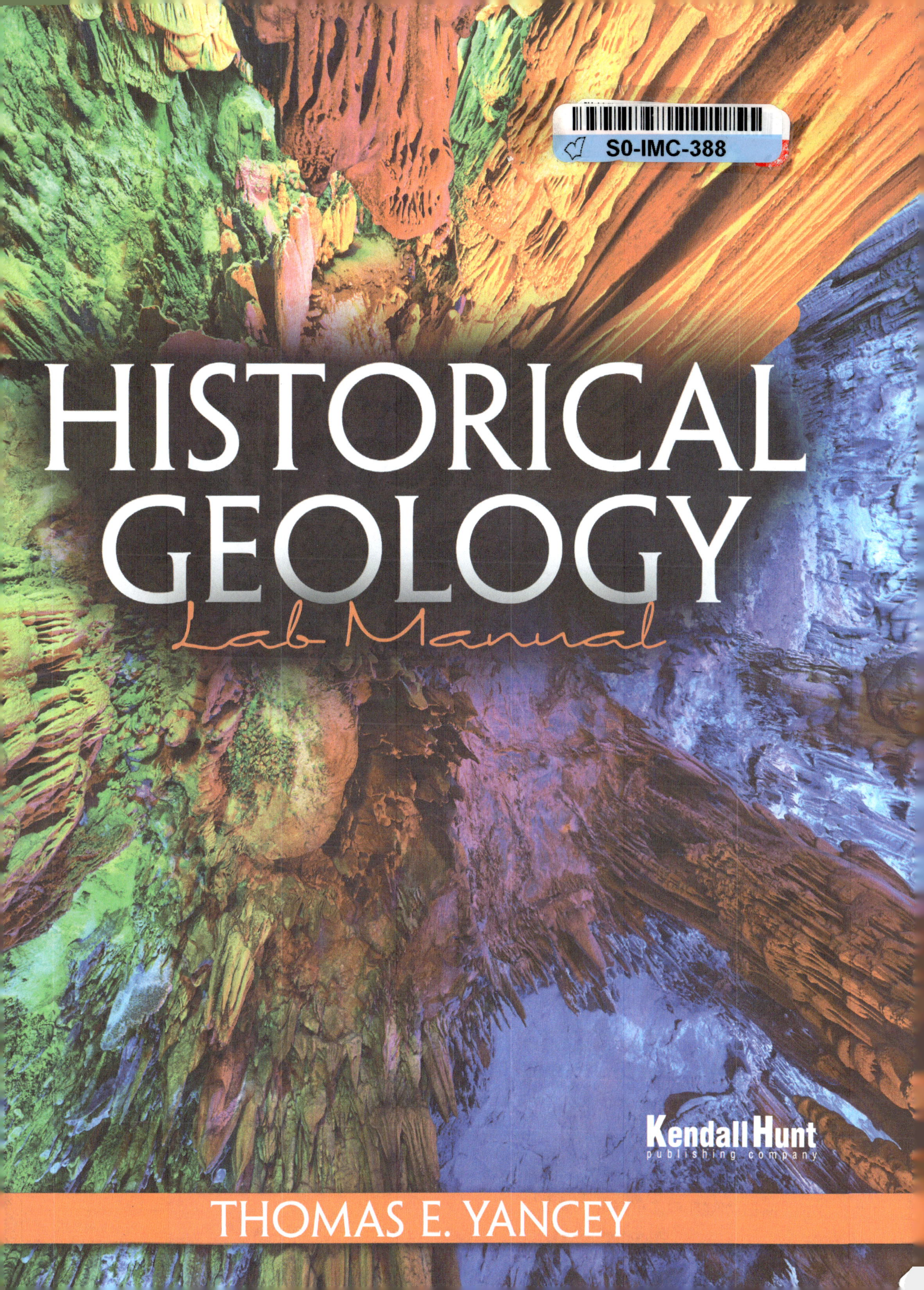

HISTORICAL GEOLOGY

Lab Manual

Kendall Hunt
publishing company

THOMAS E. YANCEY

Kendall Hunt
publishing company

www.kendallhunt.com
Send all inquiries to:
4050 Westmark Drive
Dubuque, IA 52004-1840

ISBN 978-1-4652-7465-6

Printed in the United States of America

CONTENTS

Numerical Age Determination

Lab 1

Numerical Age Dating

Numerical age dating is the determination of the number of years that have passed since a rock formed, which is most commonly done with **radioactive age dating**. This method utilizes a natural clock based on the spontaneous decay of radioactive isotopes (**parent isotopes**) to produce stable isotopes (**decay isotopes**) in a time–dependent process of radioactive decay (Table 1). The decay rate of each radioactive isotope is unique. The radioactive decay clock starts when a mineral/rock is formed. As disintegration proceeds, decay isotopes accumulate and are retained in the mineral/rock (Figures 1, 2). By measuring the amount of decay isotope in the sample and remaining parent radioactive isotope, we calculate the time elapsed since the mineral/rock was formed, using the equation

$$\text{Age} = \text{half-life} \times \log_2 [1 + \text{daughter}_{\text{now}}/\text{parent}_{\text{now}}]$$

The form of the equation is complex because the decay of a radioactive isotope follows an exponential curve. The half-life is determined by empirical lab determination and is a constant in the equation. An age calculation assumes there has been no addition or loss of parent isotope or decay isotope after the mineral/rock sample was formed. If there has been addition or loss of isotope because of chemical changes, the calculated age will be incorrect. The result must be cross-checked with independent indicators of age to test the accuracy of the calculated age. This is especially important for dating based on the decay of Potassium40, where the decay product (Argon40) is a gas that can migrate out of a mineral when it is heated and be lost.

Table 1 Commonly Used Radioactive Isotopes for Geologic Dating

Parent Isotope	Half-Life	Ultimate Stable Product	Effective Age Range
Samarium 147	106 billion years (b.y.)	Neodymium 143	> 100 million years (m.y.)
Rubidium 87	48.8 b.y.	Strontium 87	> 100 m.y.
Thorium 232	14.0 b.y.	Lead 208	> 200 m.y.
Uranium 238	4.5 b.y.	Lead 206	> 100 m.y.
Potassium 40	1.25 b.y.	Argon 40	> 100,000 years
Uranium 23S	0.70 b.y.	Lead 207	> 00 m.y.
Carbon 14	5,730 years	Nitrogen 14	0–80,000 years

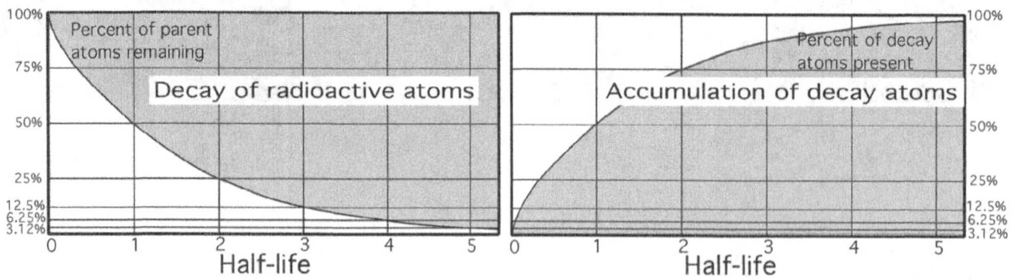

Figure 1 Exponential curves of parent and decay product for radioactive decay, showing percentage relationship to half-life. Decay of parent isotope and accumulation of decay isotope plotted against time. Time is on a linear scale, with units defined as the amount of time needed for one-half of the parent isotope to decay. This is known as the half-life. All radioactive decay follows the same exponential pattern of decay, but each radioactive isotope has a different half-life.

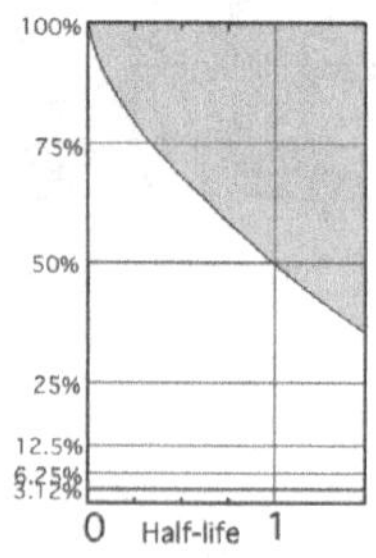

Figure 2 Decay curve for first half-life of a radiogenic isotope.

Exercises

Part 1 Knowing the half-life and knowing the amount of original radioactive isotope present in a sample, an age determination is possible by reference to charts. Assuming no inherited decay isotope and no gain or loss of isotopes to the rock from outside, answer the questions below by reference to the charts.

Procedure: The basic relationship is that half of a radioactive isotope decays during each half-life. Determine the number of half-lives passed and multiply by the length of the half-life.

1. With a 4,500 million years half-life, how old (approximately) is a sample containing a ratio of 750 atoms of ^{238}U to 250 atoms of ^{206}Pb?

2. With a 700 million years half-life, how old is a sample containing a ratio of 62 atoms of ^{235}U to 938 atoms of ^{207}Pb?

3. If the half-life of an isotope is 1,000 years and you found a specimen containing equal amounts of the parent isotope and the decay product, how old is the specimen?

4. The half-life of a radioactive isotope is 600 million years and a mineral sample containing this isotope has 1/8 the number of parent atoms it originally contained. The age of the sample is

5. If a radioactive element has a half-life of 4 million years, the amount of parent material remaining after 12 million years of decay will be what fraction of the original amount?

6. Radioactive ^{24}Na decays with a half-life of 15 hours. Starting with 1,000 atoms of this radioactive isotope, how may atoms exist after 45 hours?

7. Starting with 1,000 atoms, how many half-lives must pass to yield a mineral with 31 atoms of ^{238}U and 969 atoms of ^{206}Pb?

8. In a specimen with 12.5% parent isotope and 87.5% decay isotope, how many half-lives have passed?

9. In a sample with 75% parent isotope and 25% decay isotope, how many half-lives have passed?

10. When three-quarters (3/4) of a radioactive isotope has decayed to the stable decay isotope, how many half-lives have passed?

11. When 0.5 half-lives have passed for a mineral containing a radiogenic isotope, what percentage of the parent isotope remains in the mineral?

12. A rock contains radioactive isotopes of ^{238}U, ^{235}U, and ^{40}K. Isotopes ^{238}U and ^{235}U decay to lead (Pb), a metal that is tightly bonded with other atoms in a crystal, whereas ^{40}K decays to argon (Ar), an inert gas that does not bond in a crystal. Although minerals in the rock all formed at the same time, dating from the ^{238}U and ^{235}U indicates ages of 470 million and 468 million years, respectively, whereas the age determination from ^{40}K indicates an age of 420 million years. Which number does not give the age of rock formation?

13. Explain why one of the age determinations in the above sample is so much younger.

Dating Texas Rocks

Calculate the ages for samples of rocks collected in these Texas locations.

14. In feldspar crystals within a sample of volcanic ash collected from sediments near College Station, 50% of the original radioactive isotope remains. The isotope has a half-life of 40 million years. What is the age of the rock?

15. In feldspar crystals contained in a sample of volcanic ash collected from sedimentary strata near Dallas, 25% of the original radioactive isotope remains. The isotope has a half-life of 50 million years. What is the age of the rock?

16. In zircon crystals contained in a sample of granite collected in the vicinity of Enchanted Rock, 12.5% of the original radioactive isotope remains. The isotope has a half-life of 400 million years. What is the age of the rock?

Isochron Dating

The simple radiometric dating method described above applies to minerals and closed systems that contain radioactive isotopes and lack any inherited decay isotopes. Some are isotopes of elements that occur in many minerals and have chemical properties where other elements may easily substitute for the radioactive isotope or its decay isotope. In these cases, decay isotopes can be inherited at the time of mineral formation. The ^{87}Rb to ^{87}Sr decay is one of the best known isochron dating systems, where ^{87}Sr decay atoms may be incorporated in the mineral, substituting for Rb during mineral formation. To obtain accurate age dates, it is necessary to determine how much decay isotope was present at the beginning. This procedure, the **isochron method**, is used in all radioactive dating work and is a test for determination of a closed system. This isochron method provides a means of checking on the accuracy of calculating dates when parent isotope or decay isotope have entered or left the sample after formation.

This method utilizes an additional measurement to determine the amount of inherited decay isotope. In addition to measurement of parent and decay isotopes, measurements are made of an isotope of the decay element that is not produced by radioactive decay. **The procedure involves taking measurements from several different samples formed from the same pool of atoms**, such as an igneous rock made of several different minerals.

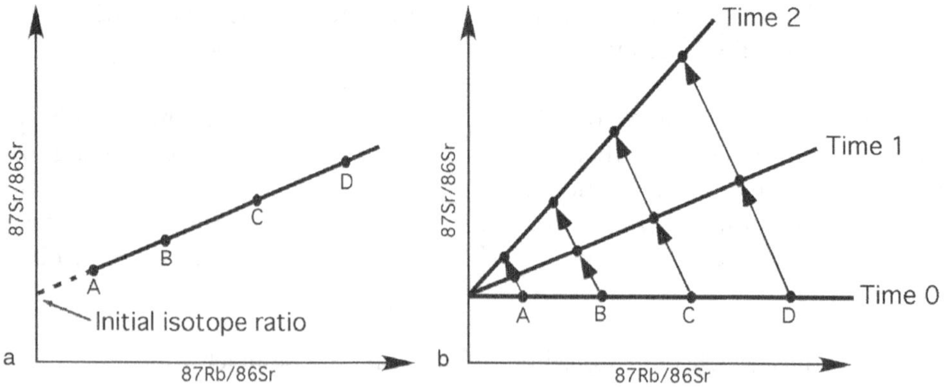

Figure 3 (a). An isochron plot of four samples from a single rock, showing the linear relationship of parent to decay isotopes for increasing concentrations of parent isotope and resulting enrichment in decay isotope. (b). Plot showing the change in slope of the isochron with increasing age of the rock sample. The angle of the slope on the isochron is a measure of the age of the rock. The initial isotope ratio of stable to unstable isotope is calculated by projection to the *y*-axis.

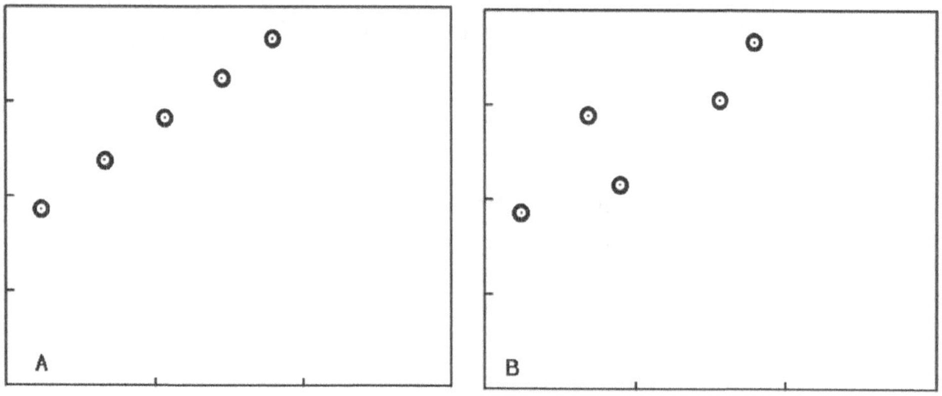

Figure 4 Plots of Rb and Sr ratios obtained from sample of the same rock mass.

In a **closed system**, ratios of the parent and decay isotope and of the decay isotope and different isotope will plot as a straight line—the **isochron** (Figure 3a). The slope of the isochron line will increase as the age of the sample increases, so measurement of the slope of this line can be used to calculate a numerical age.

If the plotted values show a scatter and fail to plot on a line, it indicates that the sample set is unreliable for age determination and will produce an incorrect age. For a linear plot, regression analysis can determine a slope of the best-fit line (used to calculate an age) and a variance in the calculated slope (amount of uncertainty in the calculated age).

17. Of the isochron plots shown in Figure 4, which can be used for calculating an accurate age?

State the reason for accepting or rejecting each plot for age determination.

A.

B.

Part 2 Using data in Table 2 of of $^{87}Sr/^{86}Sr$ and $^{87}Rb/^{86}Sr$ ratios (taken from geological reports), plot the measured values of Rb and Sr isotope ratios on the graph in Figure 5. Draw a best-fit line for the slope connecting data points to determine if they yield an isochron line for a group of samples and if a line can be drawn, extend it to the *y*-axis to calculate the initial ratio of ^{87}Sr to ^{86}Sr contained in the source magma. A hand-drawn line is adequate for this exercise, but the slope can also be determined using the graphing function in Excel. The slope corresponds to the age of the sample set and is used to calculate a numerical age. A numerically precise age can be calculated by using the values of $^{87}Sr/^{86}Sr$ and $^{87}Rb/^{86}Sr$ in the general age equation.

The equation used to calculate ages in the Rb-Sr system is

$$^{87}Sr/^{86}Sr = (^{87}Sr/^{86}Sr)_{initial} + (^{87}Rb/^{86}Sr) \cdot (e^{-\lambda t} - 1)$$

This is the mathematical expression of the best-fit line shown on your graph: Y = mX + b, where the slope is m and Y intercept is b.

Because $e^{-\lambda t} - 1 \approx \lambda t$ for $^{87}Rb/^{87}Sr$ decay, the equation can be simplified to $t \approx$ slope/ λ

where λ is the decay constant for the breakdown of ^{87}Rb to ^{87}Sr, a constant with a value of 1.42×10^{-11}

Use this form of the equation to calculate an age by inserting the measure of slope from the isochron plot: $t \approx$ (your value of slope)/(1.42×10^{-5}) $\approx$ age Ma (million years)

Calculate the slope angle with the formula: Slope = Rise/Run.

Procedure: On a graph, measure the vertical separation (rise) of the highest and lowest point and the horizonal separation (run) of the left-most and right-most point, then divide the two sums to calculate a numerical value for the slope. Calculate an age for each sample set. Note the great difference in slope angle, indicating different ages. Before doing the calculations, look over the graph and decide which sample set (A, B, or C) is the oldest and which is youngest.

Table 2 **Rb and Sr Isotope Ratios for Sample Sets from Meteorites, Grand Canyon, and Appalachian Mountains**

Sample Sets	Measured Ratio	Measured Ratio			Measured Ratio	Measured Ratio
Sample Set A	**87Rb/86Sr**	**87Sr/86Sr**		**Sample set B**	**87Rb/86Sr**	**87Sr/86Sr**
A1	0.86	0.757		B1	1.60	0.7306
A2	0.79	0.752		B2	1.32	0.7270
A3	0.72	0.747		B3	1.13	0.7240
A4	0.60	0.739		B4	2.69	0.7480
A5	0.09	0.706		B5	2.30	0.7427
				B6	0.33	0.7095
Sample Set C	**87Rb/86Sr**	**87Sr/86Sr**				
C1	7.72	0.7468				
C2	1.26	0.7111				
C3	1.78	0.7132				
C4	3.90	0.7239				
C5	5.82	0.7371				

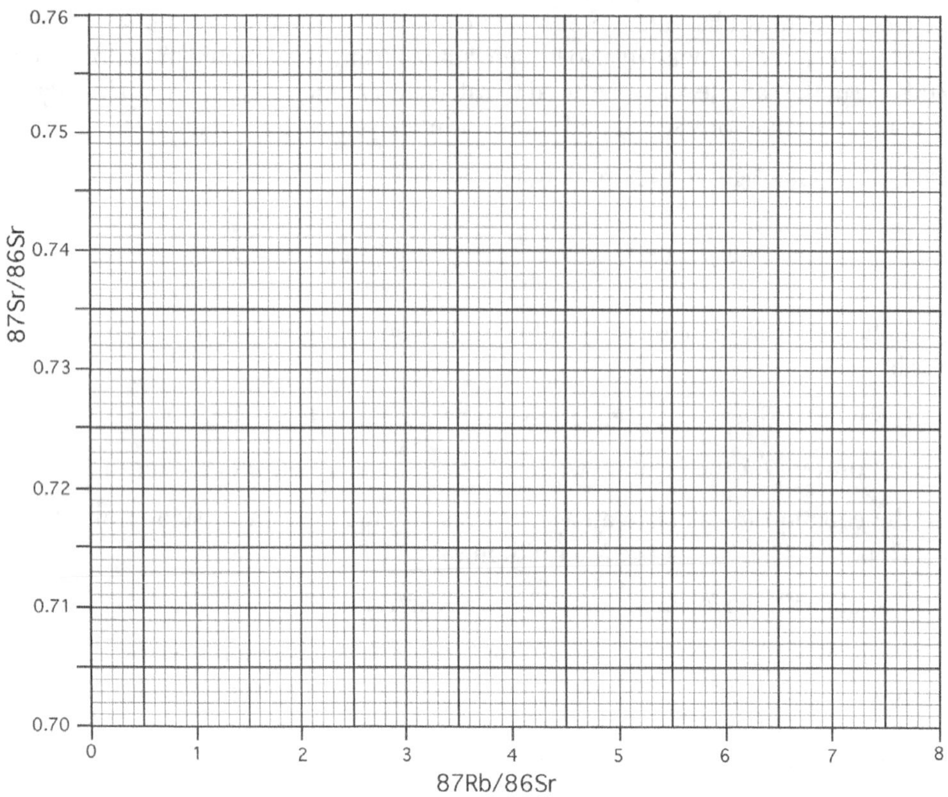

Figure 5 Graph for plotting isotope ratios used in $^{87}Rb/^{87}Sr$ dating.

18. Calculated age of sample set A:

19. Calculated age of sample set B:

20. Calculated age of sample set C:

This method is used for all radiometric dating, not just for the ^{87}Rb to ^{87}Sr decay system. This is a simple representation of the system and does not account for error in measurements. However, it is a good first approximation.

U–Pb Dating

The U–Pb system is the most robust geochronometer for dating rocks that are 100s of millions to billions of years old, because it has two completely independent "clocks." Any rock or mineral that contains uranium will contain two isotopes: ^{238}U and ^{235}U. Each isotope decays at a different rate according to its decay constant λ, which is a way of expressing the isotope's half-life. Furthermore, the ratio of the two uranium isotopes is constant in nature: any rock or mineral that contains U will have 137.88 atoms of ^{238}U for every one atom of ^{235}U.

An age can be calculated from the decay of ^{238}U to its stable decay product, ^{206}Pb, according to

$$^{206}Pb = {}^{238}U \cdot (e^{\lambda_{238}t} - 1)$$

Another, independent age can be calculated from the decay of ^{235}U to its stable decay product ^{207}Pb according to

$$^{207}Pb = {}^{235}U \cdot (e^{\lambda_{235}t} - 1)$$

A third (but not independent) age can be calculated by combining the two U–Pb decay systems:

$$\frac{^{207}\text{Pb}}{^{206}\text{Pb}} = \frac{1}{137.88} \cdot \frac{(e^{\lambda_{235}t} - 1)}{(e^{\lambda_{238}t} - 1)}$$

(note that $^{235}U/^{238}U$ can be simplified to 1/137.88)

Decay constants (λ) (and half-lives) are determined by experiment, so the only unknowns are the number of atoms of ^{207}Pb, ^{206}Pb, ^{238}U, and ^{235}U and the time elapsed since the rock or mineral formed (Table 3). These values are determined by a mass spectrometer, which provides very precise measurements of $^{206}Pb/^{238}U$, $^{207}Pb/^{235}U$, and $^{207}Pb/^{206}Pb$.

λ238 is the decay constant for the breakdown of ^{238}U to ^{206}Pb, a constant with a value of 1.551250×10^{-10}, and

λ235 is the decay constant for the breakdown of ^{235}U to ^{207}Pb, a constant with a value of 9.848500×10^{-10}.

If the ages calculated by the three isotope-ratio determinations agree with one another—such ages are called **concordant**—then the age is considered to be highly reliable. **Discordant** ages—those that do not agree with one another—suggest some type of disturbance to the decay systems and the age must interpreted with caution. These ages are compared on a diagram called a **concordia diagram** (Figure 6). In this exercise, construct a concordia diagram and interpret the U–Pb age data using that diagram.

Part 3 The isotope ratio values shown in Table 4 are for a sample set of minerals from the same rock. For the U/Pb isotope ratio values shown in Table 4, use Table 3 to estimate the age indicated by each of the U/Pb ratios by interpolation from the calculated ages.

Next, plot the U/Pb isotope ratios on the graph in Figure 7 to determine which samples are concordant or discordant. The dashed line is the curve along which samples should plot if they have remained in a closed system.

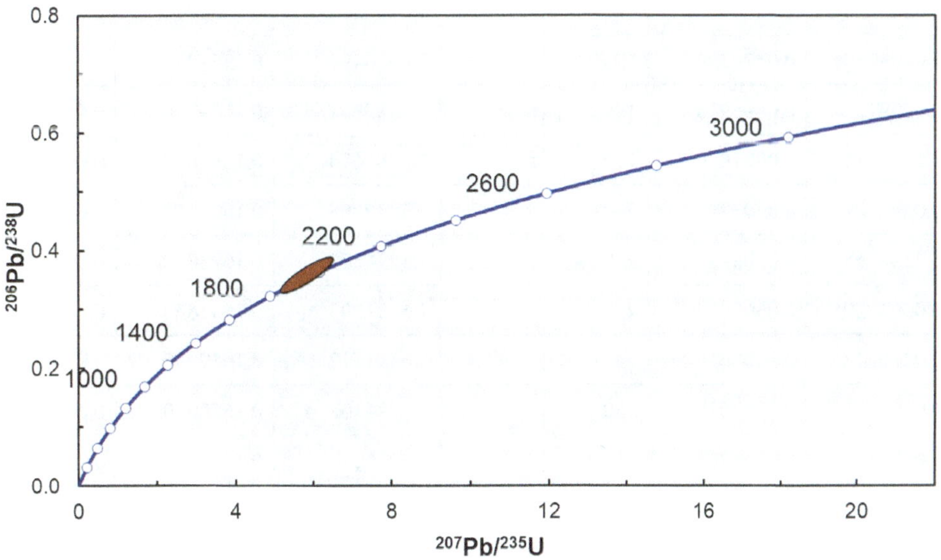

Figure 6 Example of a concordia diagram showing the placement of sample ages on the curved line formed by plotting the ratios of two radiogenic decay systems with different half lives. Dates are plotted as ovals instead of points to show the standard deviation of error involved in each measurement.

Table 3 Standard $^{206}Pb/^{238}U$, $^{207}Pb/^{235}U$, and $^{207}Pb/^{206}Pb$ Ratios for Decay Ages from 0 to 1,500 Million Years

207Pb/ 235U	206Pb/ 238U	207Pb/ 206Pb	Age (m.y.)		207Pb/ 235U	206Pb/ 238U	207Pb/ 206Pb	Age (m.y.)
0.0000000	0.0000000	-	0		1.0930817	0.1233820	0.0642539	750
0.0098972	0.0015525	0.0462371	10		1.1137973	0.1251260	0.0645591	760
0.0198923	0.0031073	0.0464299	20		1.1347179	0.1268727	0.0648662	770
0.0299863	0.0046646	0.0466238	30		1.1558455	0.1286221	0.0651753	780
0.0401802	0.0062243	0.0468189	40		1.1771823	0.1303742	0.0654863	790
0.0504751	0.0077864	0.0470152	50		1.1987302	0.1321291	0.0657994	800
0.0608718	0.0093509	0.0472127	60		1.2204913	0.1338867	0.0661144	810
0.0713714	0.0109179	0.0474114	70		1.2424679	0.1356470	0.0664315	820
0.0819749	0.0124873	0.0476113	80		1.2646620	0.1374100	0.0667505	830
0.0926834	0.0140592	0.0478125	90		1.2870757	0.1391758	0.0670717	840
0.1034979	0.0156334	0.0480148	100		1.3097112	0.1409443	0.0673949	850
0.1144193	0.0172102	0.0482184	110		1.3325708	0.1427156	0.0677201	860
0.1254489	0.0187893	0.0484233	120		1.3556566	0.1444896	0.0680475	870
0.1365877	0.0203710	0.0486294	130		1.3789709	0.1462663	0.0683769	880
0.1478367	0.0219550	0.0488367	140		1.4025159	0.1480459	0.0687085	890
0.1591970	0.0235416	0.0490454	150		1.4262940	0.1498282	0.0690422	900
0.1706697	0.0251306	0.0492553	160		1.4503074	0.1516132	0.0693780	910
0.1822560	0.0267220	0.0494665	170		1.4745585	0.1534010	0.0697160	920
0.1939570	0.0283160	0.0496790	180		1.4990496	0.1551916	0.0700562	930
0.2057738	0.0299124	0.0498928	190		1.5237831	0.1569850	0.0703985	940
0.2177075	0.0315113	0.0501079	200		1.5487614	0.1587812	0.0707431	950
0.2297594	0.0331127	0.0503243	210		1.5739868	0.1605801	0.0710899	960
0.2419305	0.0347165	0.0505421	220		1.5994620	0.1623819	0.0714390	970
0.2542221	0.0363229	0.0507612	230		1.6251893	0.1641864	0.0717902	980
0.2666353	0.0379317	0.0509816	240		1.6511712	0.1659938	0.0721438	990
0.2791714	0.0395431	0.0512035	250		1.6774102	0.1678039	0.0724996	1000
0.2918315	0.0411569	0.0514267	260		1.7039090	0.1696169	0.0728578	1010
0.3046170	0.0427732	0.0516512	270		1.7306700	0.1714327	0.0732183	1020
0.3175290	0.0443921	0.0518772	280		1.7576959	0.1732513	0.0735811	1030
0.3305688	0.0460135	0.0521045	290		1.7849892	0.1750727	0.0739462	1040
0.3437376	0.0476374	0.0523333	300		1.8125527	0.1768969	0.0743137	1050
0.3570368	0.0492638	0.0525635	310		1.8403890	0.1787240	0.0746836	1060
0.3704676	0.0508927	0.0527951	320		1.8685007	0.1805539	0.0750559	1070

207Pb/ 235U	206Pb/ 238U	207Pb/ 206Pb	Age (m.y.)		207Pb/ 235U	206Pb/ 238U	207Pb/ 206Pb	Age (m.y.)
0.3840313	0.0525242	0.0530281	330		1.8968907	0.1823867	0.0754307	1080
0.3977293	0.0541582	0.0532626	340		1.9255617	0.1842223	0.0758078	1090
0.4115629	0.0557947	0.0534985	350		1.9545165	0.1860607	0.0761874	1100
0.4255333	0.0574338	0.0537360	360		1.9837578	0.1879020	0.0765695	1110
0.4396420	0.0590754	0.0539748	370		2.0132885	0.1897462	0.0769541	1120
0.4538904	0.0607196	0.0542152	380		2.0431115	0.1915932	0.0773412	1130
0.4682798	0.0623663	0.0544571	390		2.0732296	0.1934431	0.0777308	1140
0.4828116	0.0640155	0.0547005	400		2.1036458	0.1952959	0.0781229	1150
0.4974872	0.0656674	0.0549454	410		2.1343631	0.1971515	0.0785176	1160
0.5123081	0.0673218	0.0551918	420		2.1653844	0.1990100	0.0789149	1170
0.5272756	0.0689787	0.0554397	430		2.1967127	0.2008714	0.0793147	1180
0.5423913	0.0706383	0.0556892	440		2.2283511	0.2027357	0.0797172	1190
0.5576566	0.0723004	0.0559403	450		2.2603026	0.2046029	0.0801223	1200
0.5730730	0.0739651	0.0561929	460		2.2925703	0.2064730	0.0805301	1210
0.5886419	0.0756324	0.0564472	470		2.3251574	0.2083460	0.0809405	1220
0.6043649	0.0773023	0.0567030	480		2.3580670	0.2102219	0.0813536	1230
0.6202436	0.0789747	0.0569604	490		2.3913023	0.2121007	0.0817694	1240
0.6362794	0.0806498	0.0572194	500		2.4248665	0.2139825	0.0821880	1250
0.6524739	0.0823274	0.0574801	510		2.4587630	0.2158671	0.0826093	1260
0.6688287	0.0840077	0.0577424	520		2.4929949	0.2177547	0.0830334	1270
0.6853454	0.0856906	0.0580063	530		2.5275656	0.2196452	0.0834602	1280
0.7020255	0.0873760	0.0582719	540		2.5624785	0.2215386	0.0838899	1290
0.7188707	0.0890642	0.0585392	550		2.5977369	0.2234350	0.0843223	1300
0.7358826	0.0907549	0.0588081	560		2.6333442	0.2253344	0.0847577	1310
0.7530629	0.0924482	0.0590788	570		2.6693040	0.2272366	0.0851958	1320
0.7704133	0.0941442	0.0593511	580		2.7056197	0.2291419	0.0856369	1330
0.7879353	0.0958428	0.0596252	590		2.7422948	0.2310500	0.0860809	1340
0.8056308	0.0975440	0.0599010	600		2.7793329	0.2329612	0.0865278	1350
0.8235014	0.0992479	0.0601785	610		2.8167375	0.2348753	0.0869777	1360
0.8415489	0.1009545	0.0604578	620		2.8545123	0.2367924	0.0874305	1370
0.8597750	0.1026636	0.0607389	630		2.8926611	0.2387125	0.0878863	1380
0.8781815	0.1043755	0.0610217	640		2.9311873	0.2406355	0.0883451	1390
0.8967701	0.1060900	0.0613064	650		2.9700949	0.2425615	0.0888070	1400
0.9155427	0.1078071	0.0615928	660		3.0093876	0.2444906	0.0892719	1410

(Continued)

Table 3 Standard $^{206}Pb/^{238}U$, $^{207}Pb/^{235}U$, and $^{207}Pb/^{206}Pb$ Ratios for Decay Ages from 0 to 1,500 Million Years (*Continued*)

207Pb/ 235U	206Pb/ 238U	207Pb/ 206Pb	Age (m.y.)		207Pb/ 235U	206Pb/ 238U	207Pb/ 206Pb	Age (m.y.)
0.9345012	0.1095269	0.0618810	670		3.0490691	0.2464226	0.0897399	1420
0.9536472	0.1112494	0.0621711	680		3.0891434	0.2483576	0.0902110	1430
0.9729828	0.1129746	0.0624630	690		3.1296142	0.2502956	0.0906852	1440
0.9925097	0.1147024	0.0627568	700		3.1704857	0.2522366	0.0911625	1450
1.0122299	0.1164330	0.0630524	710		3.2117616	0.2541807	0.0916430	1460
1.0321452	0.1181662	0.0633500	720		3.2534461	0.2561277	0.0921268	1470
1.0522577	0.1199021	0.0636494	730		3.2955431	0.2580778	0.0926137	1480
1.0725692	0.1216407	0.0639507	740		3.3380568	0.2600309	0.0931038	1490
1.0930817	0.1233820	0.0642539	750		3.3809912	0.2619870	0.0935972	1500

Table 4 Measured Isotope Ratios for a Group of Samples

	Measured Ratio	Measured Ratio	Measured Ratio	207Pb/ 235U	206Pb/ 238U	207Pb/ 206Pb	Concordant or
Sample	**207Pb/ 235U**	**206Pb/ 238U**	**207Pb/ 206Pb**	**Age (Ma)**	**Age (Ma)**	**Age (Ma)**	**Discordant?**
D1	2.1810100	0.1999404	0.0791145				
D2	2.1186305	0.1943691	0.0790545				
D3	2.1778787	0.1997543	0.0790745				
D4	1.6733447	0.1534010	0.0791145				
D5	2.1794440	0.1998473	0.0790945				
D6	1.3547112	0.1242536	0.0790745				
D7	2.1841443	0.2001265	0.0791545				
D8	2.1763142	0.1996612	0.0790545				

21. Indicate whether each point is concordant or discordant.

22. Which samples provide reliable age dates and which samples do not?

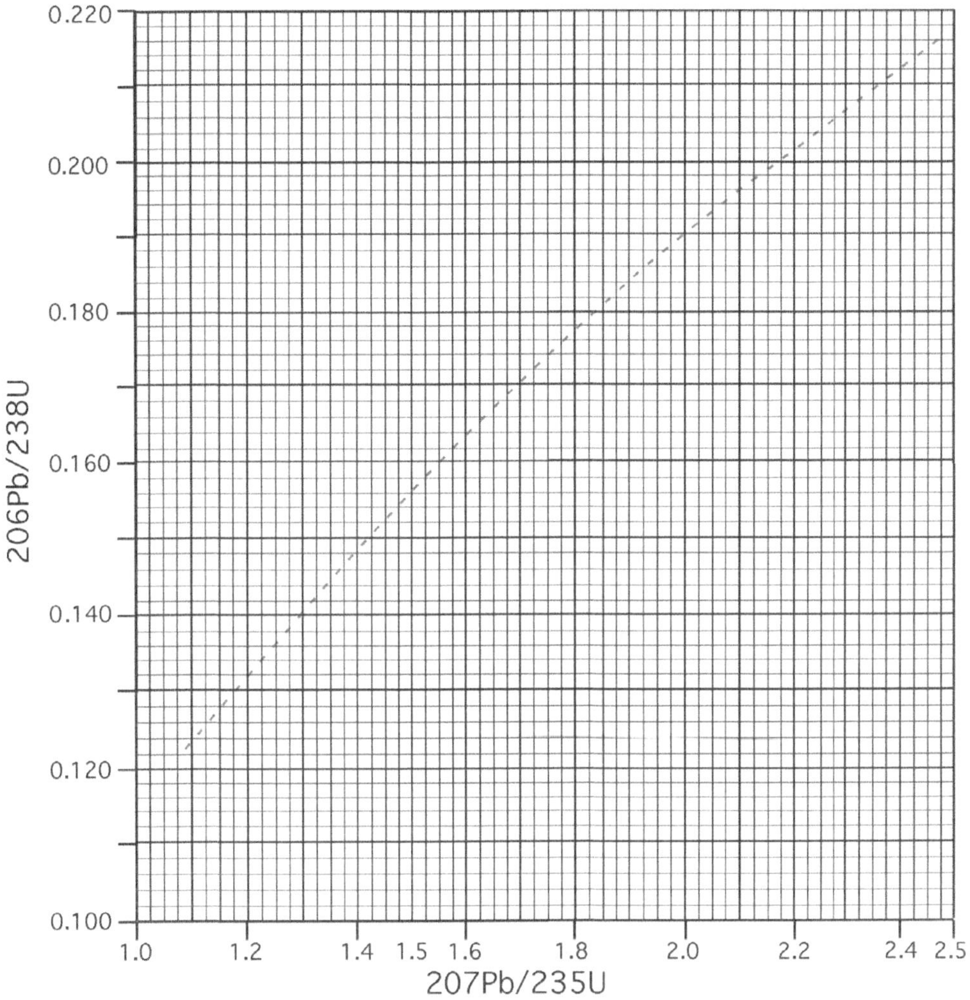

Figure 7 Graph for plotting isotope ratios used in U–Pb dating.

Siliciclastic Sediments and Sedimentary Rocks

Lab 2

Introduction

Sediment refers to all solid particles transported and deposited by water, wind, glaciers, gravity, and solids precipitated at the Earth surface. Most sediment is produced from the breakdown of igneous and metamorphic rocks that form at temperatures and pressures much higher than those at the Earth's surface and in an environment containing less water, oxygen, carbon dioxide, and organics. These rocks are unstable under atmospheric conditions and undergo chemical and physical changes when brought to the surface and **weathered**. Detrital sediment consists of mineral and rock fragments.

Rock Fragments/Chert

Most large detrital grains are composed of more than one crystal (polycrystalline). Cobbles and boulders can be identified as pieces of volcanic, metamorphic/plutonic, or sedimentary rock to provide information on the source area. Smaller polycrystalline grains are just called rock fragments. A common polycrystalline grain in sediments of our area is microcrystalline silica: **chert**. Chert occurs commonly in limestone as a replacement material (it is a silica concretion). Chert concretions are hard and uniform in composition, making them excellent material for manufacturing arrowheads, spear points, etc. Central Texas limestones contain chert concretions that weather out and become concentrated on the ground and are common in Texas rivers. **Learn to recognize chert**.

The **primary textural characters** of sediment particles are the **1) size of grains, 2) sorting of grains in a sediment, and 3) grain roundness and sphericity**.

Sediment Grain Size

Sediment grain size is an important indicator of energy in the environment of deposition. Fine sediment (clay and silt) is deposited from quiet water, sand sediment is moved by flowing water, waves and wind, and coarse (gravel and larger) requires high energy to be moved. These provide important indications of depositional conditions. The Wentworth grain size scale is used to describe sediment grain size (Figure 1). Note that the clay grain size category refers to particles 4 micrometers and smaller and refers to all particles of that size, whether or not they are members of the clay mineral group.

Size (mm)	Clast	Sediment	Rock
	Boulder		
256	Cobble	gravel	Conglomerate
64	Pebble		
4	Granule		
2	Sand	Coarse sand Medium sand Fine sand	Sandstone
1/16	Silt	mud	Mudstone-Shale
1/256	Clay		

Figure 1 Wentworth grain size scal.
From Wentworth, Chester K., "A Scale of Grade and Class Terms for Clastic Sediments," *The Journal of Geology*, Vol. 30, No. 5 (Jul.–Aug., 1922), p. 381, Table 1, 1922.

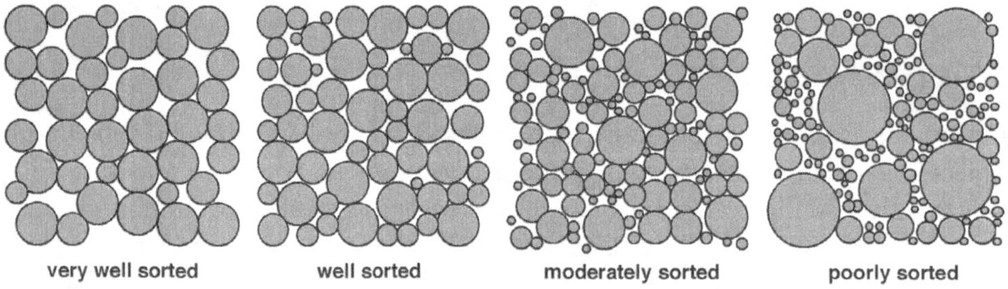

Figure 2 Standard sorting chart.

Sorting

Sediment forms as mixture of different sized grains but during transport and deposition it is **sorted** into sizes by the energy present in an environment. The degree of sorting (Figure 2) varies with the depositional environment. Sediment having a narrow range of sizes is called "well-sorted."

Roundness and Sphericity

Texture also includes the shape of individual particles: their roundness and sphericity (Figure 3). Roundness is the extent to which corners on a clast have been smoothed and rounded, in contrast to sharp, angular corners. Sphericity is the extent that the grain or clast approaches a spherical form, whether or not it has angular or

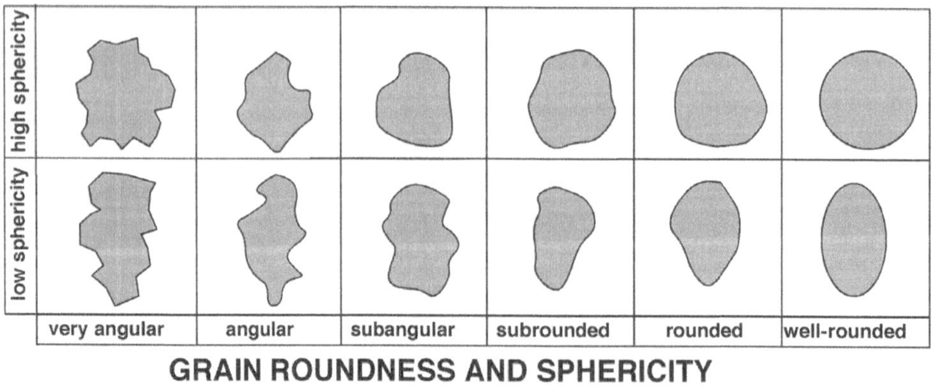

Figure 3 Roundness and sphericity of grains.

sharp corners. For a clast that starts with an angular, non-spherical form, abrasion will first tend to round off the corners of the grain and then produce a more spherical form as the size of the clast is reduced. A grain can have a high degree of roundness but a very low sphericity but it is unlikely for a grain to have a high sphericity and a low degree of roundness.

Classification of Siliciclastic Sediments and Rocks

Siliciclastic sediments are classified by grain size and mineral composition. The first level of determination is grain size. If it is **coarse, it is gravel (for sediment) or conglomerate (for rock)**;

if it is ~**millimeter size, it is sand (for sediment) or sandstone (for rock)**;

if it is **very fine, it is mud (for sediment) or mudstone/shale (for rock)**. See Figure 1 for size limits.

Composition of sand size siliciclastic sediment is determined by the percentage of three categories of grains: **quartz, feldspar (all varieties), and poly-crystalline grains (rock fragments)**. These are plotted on a ternary diagram (Figure 4), which has three scales of 0%–100% along the three sides of the diagram. Because the mineral quartz is dominant in most sands, for a sand to be called as **quartzose sand** it must be composed of 90% or more quartz. If the non-quartz faction contains more feldspar grains than rock fragment grains it is called a **feldspathic sandstone** (or arkose) and if it contains more rock fragment grains than feldspar grains it is called a **lithic sandstone.**

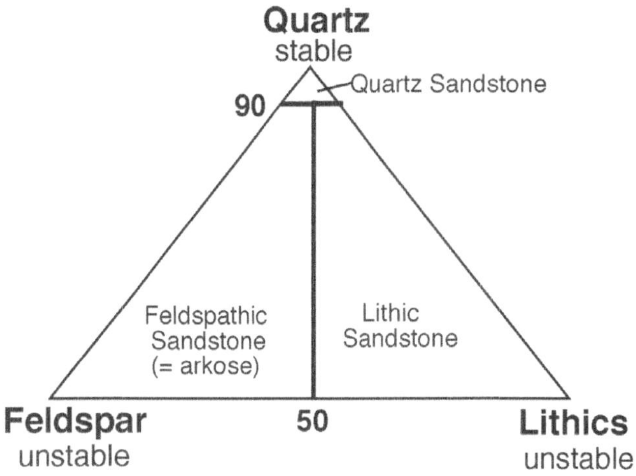

Figure 4 Classification of siliciclastic sediments.

Siliciclastic sediment can contain other clasts, like volcanic ash or skeletal remains, or precipitates from pore fluids (cement). If the other material is dominant, it is named using a different sediment classification.

PreLab Exercise

To understand the origin, weathering, and transport history of sediment derived from outcrops, the composition of minerals forming sediment must be determined (see Figure 5). Silicate mineral groups are determined by the arrangement of silica tetrahedra ions in the crystal structure. Review the silicate minerals that form the great majority of crustal rocks before starting work on sediments. Some useful characters of silicate minerals are

	Color	Cleavage	Mohs Hardness	Crystal Structure
Olivine	green	none	7	single tetrahedron
Pyroxene	dark	2 cleavage angles	5–6	chain silicate
Amphibole	dark	2 cleavage angles	5–6	chain silicate
Mica	shiny	platy cleavage	2–3	sheet silicate
Feldspar	light	3 cleavage angles	6	3D tetrahedra array
Quartz	white, clear	none	7	3D tetrahedra array

Figure 5 Minerals of the basic rock-forming silicate mineral groups. (Note: for olivine, the olivine-rich rock is a sample of mantle peridotite brought up to the surface by a volcano; the dark edges are basalt that contained the peridotite block.)

1. List one or two characters useful for recognition of the following minerals. Turn in a copy of this list to your Teaching Assistant at the start of the lab period.

   A. Pyroxenes and amphiboles: ______________________, ______________________

   B. Mica: ______________________, ______________________

   C. Feldspar: ______________________, ______________________

   D. Quartz: ______________________, ______________________

2. List differences between **feldspar** and **quartz** that allow you to recognize one from the other.

Exercise Work

3. You have 5 pebbles in a sample from a river gravel transported far from a source. To determine the roundness and sphericity, measure the pebbles (in mm) in the broadest 2 dimensions, record the data and determine sphericity.

   Ratio of ~1:1 or close: **high** sphericity
   1:1.25 to 1:1.75: **moderate** sphericity
   1:1.75 or greater: **low** sphericity

   Record the measurements on the pebbles and calculate sphericity. Estimate the degree of roundness on corners, using the roundness classes in Figure 3.

	#1	#2	#3	#4	#5
maximum (mm)					
minimum (mm)					
sphericity					
roundness					

Which character (roundness or sphericity) is a better indicator of transport? Explain your answer, using the ideas that sphericity is mostly produced by breakage and roundness is mostly the result of abrasion.

4. Compare samples of crystalline **quartz** and **chert** (microcrystalline quartz), then list differences for identifying one from the other.

5. You have four sediment samples to examine: A, B, C, and D. These samples have different grain sizes, mineral composition, degree of sorting, and grain roundness. Note the variation among particles in the sample of grains.

   Arrange the sediment samples according to grain size, from the coarsest to finest. Use mean grain size.

   Coarse **Fine**

   __________ __________ __________ __________

6. Arrange the sediment samples according to sorting of the sediment, from the best sorted to most poorly sorted.

   Well-sorted **Poorly-sorted**

   __________ __________ __________ __________

7. Arrange the sediment samples A, C, and D according to the grain roundness, from the most round to the most angular. **Use quartz grains for this determination** and examine particles that are of average grain size or larger.

   Angular **Round**

   __________ __________ __________

8. Using a larger sample of pebbles, sort them into groups of different composition (mineral/rock type). In this sample, you have mostly clasts of 1) quartz, 2) chert, and 3) rock fragments. Work with your Teaching Assistant to verify composition.

9. You will be given a photo of a sample of sand grains. Make a visual estimate of roundness and sphericity of the 10 marked sand grains, then estimate the average roundness and sphericity of the grains. Based on all grains in the photo, determine sorting of the sediment.

   Average Roundness: __________ Average Sphericity __________ Sorting __________

10. Use a binocular microscope to determine the average roundness of sand grains in sediment labeled St. Peter Formation sand and Simsboro Formation sand.

    A. Simsboro sand __________ St. Peter sand __________

    B. The St. Peter sand is the preferred sand for use in fracing (fracturing) wells during oil and gas production. Based on the grain characters, explain why St. Peter sand is better frac sand than Simsboro sand.

11. Use a binocular microscope and grain size scale to determine the average grain size of coarse sediment from Enchanted Rock (Figure 6).

What is the mean grain size? _________ What is the sorting? __________

What is the average roundness? __________ What is the parent rock type that produced this sediment. __________

Determine the composition of 30 sediment grains. Use the listed criteria and grain photos to identify grains; list number of each type of grain:

		Number of grains
White, no cleavage —	Quartz	__________
Pink, cleavage faces —	Feldspar	__________
Polycrystalline —	Rock fragments	__________
Black, irregular —	Hornblende	__________
Black, flat cleavaage face —	Biotite mica	__________

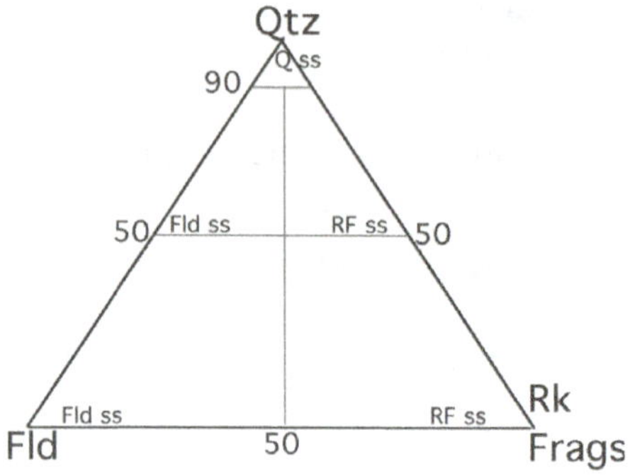

Plot the values of quartz, feldspar, and rock frags on the ternary diagram. Ignore the hornblende and biotite.

If the results are unrepresentative of the sample, repeat the count until you get a reasonable number of grains to plot.

12. Use a binocular microscope and grain size scale to determine the average grain size of coarse sediment from a beach at Oceanside, California (Figure 7).

What is the mean grain size? _________ What is the sorting? __________ What is the average roundness? __________

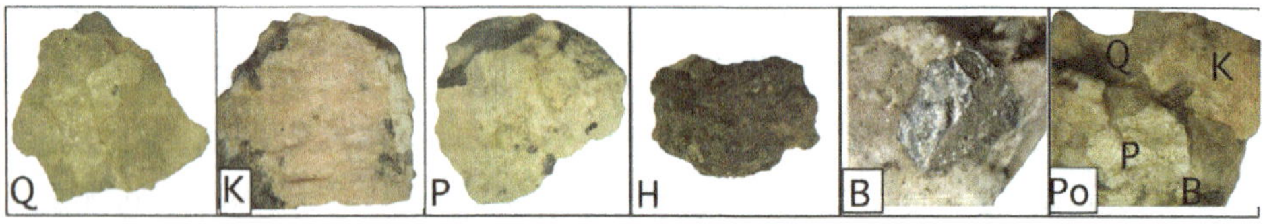

Figure 6 Enchanted Rock sediment grain types: quartz (Q), K feldspar (K), plagioclase (P), hornblende (H), biotite (B), and polycrystalline (Po), with quartz, K feldspar, plagioclase, and hornblende marked on grain.

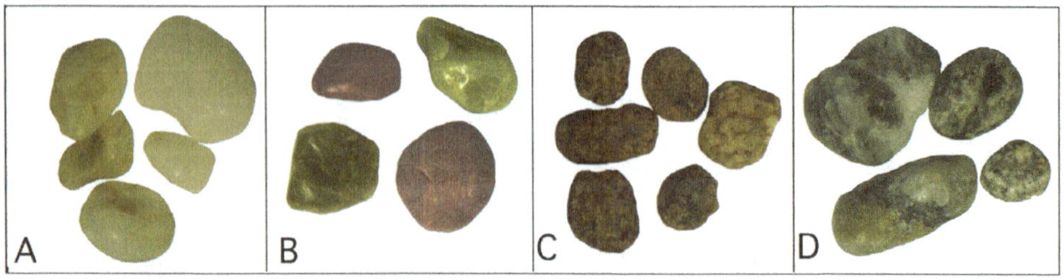

Figure 7 Oceanside sediment grain types: A) quartz, B) chert, C) sandstone, and D) polycrystalline rock fragments.

Determine the composition of 30 sediment grains. Use the listed criteria and grain photos to identify grains:

		Number of grains
White —	Quartz	
Red or gray-green —	Chert	
Polycrystalline —	Rock fragments	
Fine granular polycrystalline —	Sandstone	
Indeterminate —	other	

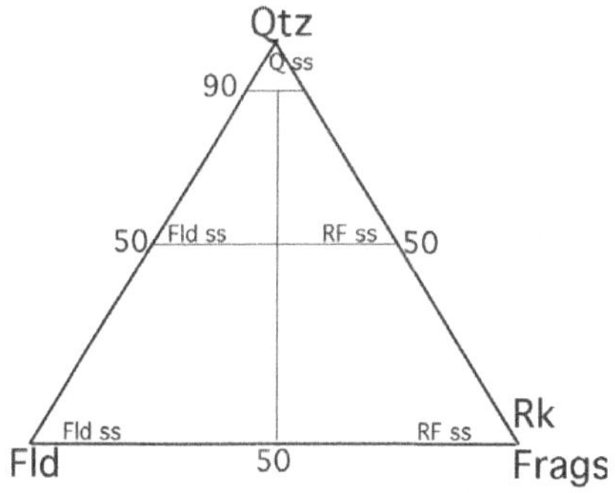

Plot the values. **Combine counts of quartz and chert into a single value of Qtz (quartz) and combine counts of rock fragments and sandstone into a single value of Rk Frags (rock fragments)** and plot on the ternary diagram to the right. Ignore the indeterminate grains. There is no feldspar. **If the results are unrepresentative of the sample, repeat the count until you get a reasonable number of grains to plot.**

13. Use a binocular microscope and grain size scale to determine the average grain size of loose sand samples. Determine grain size as fine sand, medium sand, or coarse sand size. Determine the degree of sorting and roundness using sorting and roundness comparators in Figures 2 and 3.

	Sediment Name	Grain size	Sorting	Av. Roundness
Sediment # EE				
# HH				
# JJ				

14. Compare samples of an igneous-metamorphic source rock with a sediment (#52) derived from that rock. List 2 or more differences that distinguish the sediment rock from the source rock.

A.

B.

15. Determine the rock type (name) of each sample, using one of the rock names given in the the siliciclastic rock classification presented above. List the major mineral and the secondary minerals present in each rock. Use the flowchart (Figure 10) to help identify sediments and rocks.

	Rock Name	Major mineral	Secondary minerals (if determinable)	Sorting
#34				
#35				
#41				
#45				
#55				

(Continued)

	Rock Name	Major mineral	Secondary minerals (if determinable)	Sorting
#64				
#72				
#75				
#79				

Describing a Sediment Core

16. Using **sediment peels** of modern sands in a tidal pass environment at San Luis Pass, west end of Galveston Island, divide the core into units of similar sediment and name the sediment type (Figure 8). For this task, use the modifier terms "shelly" and "muddy" to better describe the sediment. Idenify sediment name and the sediment sorting in each section.

Procedure for Describing Core

A. Divide the core peel into units by selecting boundaries on **intervals of same grain size, zones of shell content, or presence of thin muds along the core.**

B. **Record the level of change of sand units** on the thickness scale below (Figure 9).

C. For sand units defined by sediment change boundaries, **fill the units with symbols** for the type of sediment present.

D. Where the sand shows thin layering (lamination), draw some thin lines in the unit in addition to sediment symbol (Figure 8).

E. Write the **sediment name and sorting** on the right side of the page for each unit.

Figure 8 Legend for core description.

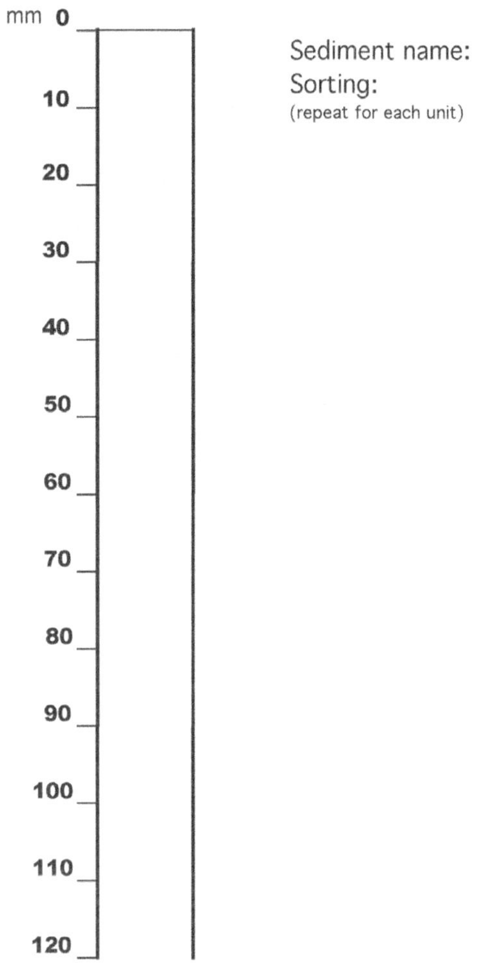

Figure 9 Thickness scale for naming sediments in San Luis Pass sediment peels.

FLOWCHART FOR IDENTIFYING SEDIMENTARY ROCKS

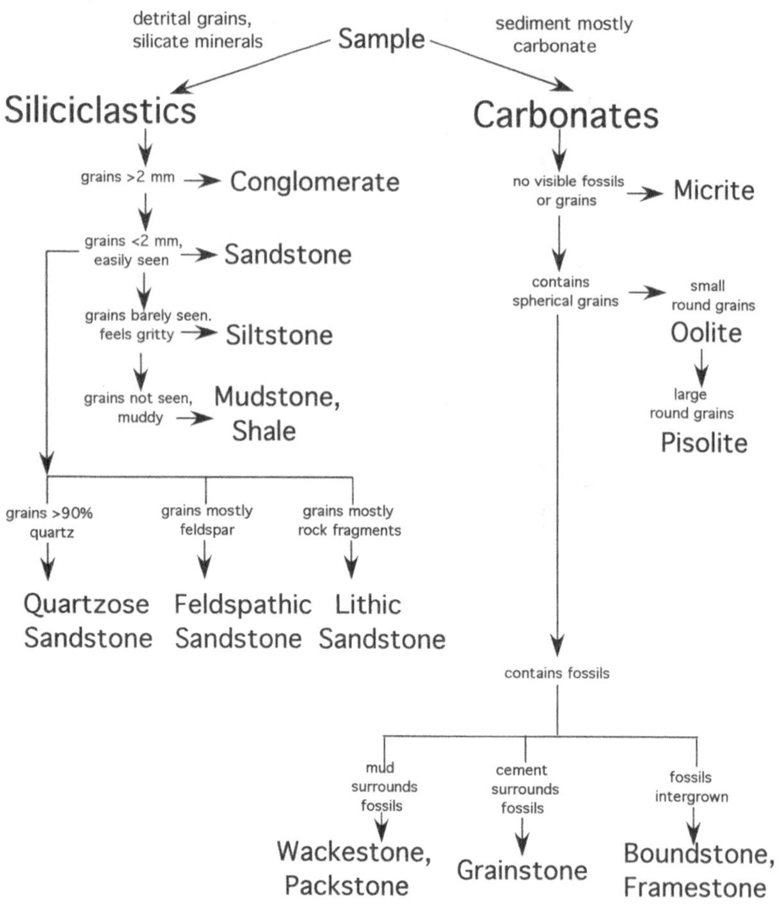

Figure 10 Flowchart for identifying sediments and sedimentary rocks.

Carbonate Sediments and Sedimentary Rocks

Lab 3

Introduction

Carbonate sediment is different from siliciclastic sediment by being produced in the depositional environment rather than being detrital pieces produced by weathering of rock. Most carbonates are biochemical materials secreted as hard skeletal parts made by organisms. These skeletal parts become carbonate grains when the organism dies and becomes bioclasts. The main difference from siliciclastic grains is that carbonate grains are softer and more soluble. Thus, carbonate grains are more likely than other minerals to recrystallize, to be dissolved, and reprecipitate as cement.

A sediment with more than 50% carbonate content is a carbonate sediment.

Mineral Group	Minerals
Carbonate group	Calcite, aragonite, dolomite

The common carbonate **minerals** are **calcite, aragonite**, and **dolomite** and all are members of the carbonate mineral group. Calcite and aragonite are the minerals secreted by organisms, whereas dolomite occurs as a replacement mineral by the addition of Mg ions to calcite or aragonite. Aragonite is the least stable carbonate mineral and inverts to calcite over time. It is rare in sediments over 100 million years old. Over longer time intervals, calcite will commonly be replaced by dolomite.

Classification of Carbonate Sediments and Rocks

Grain types are as follows:

bioclasts—shells or pieces of shell

ooids—small grains with smooth concentric layers, formed when water precipitates carbonate under conditions of regular agitation

oncoids (oncoliths)—large carbonate masses with irregular concentric layers, formed both by algal films and precipitation, sometimes by calcareous algae

intraclasts—pieces of carbonate mud broken up and moved by storm waves or strong water currents

Carbonate grains in a rock are held together by

micrite—carbonate mud produced by organisms or as a precipitate

cement—precipitated from pore fluids after the grains are deposited

Carbonate sediments and rocks are classified by grain type and texture (Figure 1).

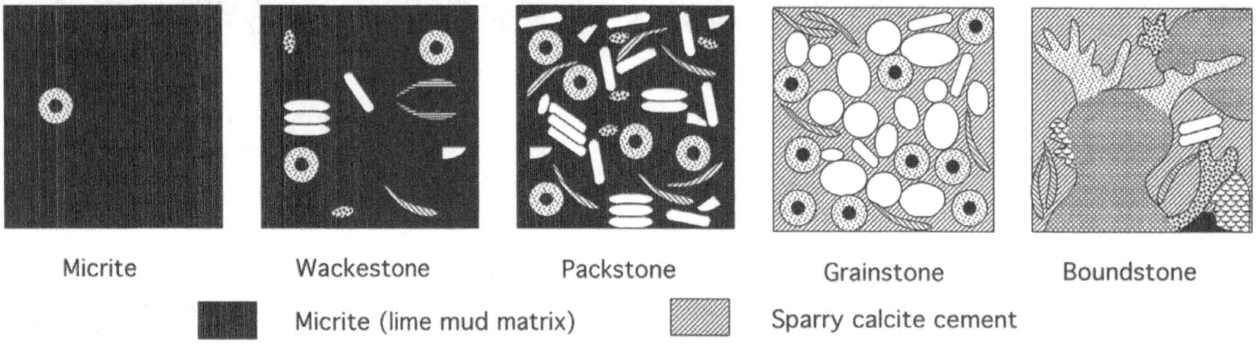

Figure 1 Grain—textural classification of carbonate sediments.

Carbonate Classification (Dunham Classification)

Micrite	—composed of carbonate mud (grain content may range up to 10%)
Wackestone–Packstone	—a mixture of bioclasts and carbonate mud with subdivisions of wackestone: mud-supported; more carbonate mud than bioclasts (<50% bioclasts) packstone: grain-supported; more carbonate mud between grains (>50% bioclasts)
Grainstone	—carbonate grains (bioclasts and other types), with carbonate cement filling spaces between grains (grain-supported)
Boundstone (Framestone)	—composed of intergrown skeletal material (reef rock)

Wackestone–Packstone: A standard carbonate classification distinguishes between wackestone and packstone, but it is difficult to consistently distinguish them and it is not important to make the distinction for this exercise, so lump these two rock types together as wackestone–packstone.

Micrite mud versus micrite rock. The name **micrite** has two meanings: it means carbonate mud composed of tiny crystals/needles produced by organisms and/or released by the breakup of skeletons, and it is the name of the cemented rock composed entirely or mostly of micrite mud.

Ooids and Oncoids (Oncoliths)

Ooids are spherical grains formed by grains that are frequently moved and have carbonate precipitate or accrete onto their outer surface. They grow like a plant by adding more material onto the outer surface. Most of this occurs with the aid of a film of microbes (bacteria, algae, etc.) that allow particles to be pasted onto the surface or promote precipitation within the microbial film. **Oncoids (oncoliths)** are larger (cm or dm size), more irregular rounded particles usually forming with knob protrusions where growth occurred faster than adjacent portions. Some calcareous algae that secrete a carbonate skeleton grow in this manner and create very large algal balls in deep water.

Intraclasts

Intraclasts are small, flat slabs of carbonate mud broken out of the sediment surface and moved. Carbonate cementation starts very early in seafloor sediments, so high-energy current or wave disturbance breaks crusts into thin pieces. Intraclasts are most common in ancient carbonates.

Additional carbonate rock names for distinctive deposits:

Chalk—sediment composed of tiny, microscopic skeletal particles produced by plankton

Oolite—sediment composed of ooids and/or pisoliths (large ooids)

Coquina—a cemented mass of rounded shell clasts formed on a beach

Rudstone—a deposit of large blocks of reef rock (boundstone/framestone) shed from the steep front of growing reefs, where blocks break loose and fall into deeper water.

Cementation

The transformation of sediment into a sedimentary rock is lithification, a process resulting from **cementation** of grains in the sediment. Dissolved compounds precipitate in pores between grains and in sediments containing clay or carbonate mud, the fine materials present recrystallize to produce cement (like the setting of concrete) surrounding the large grains. Cementation occurs during **diagenesis**, the name for a group of chemical changes that occur under conditions of shallow burial of sediments and rocks. Common cements include calcite, silica, and clays. Cementation can be very local and produce **concretions** in the sediment.

Exercise Work

1. Examine specimens of the decorative building stone called Texas rock (A9). It is limestone that contained many fossils, but the fossils are dissolved out whereas the matrix remains. The matrix is calcite; the fossil shells were aragonite. Both have the same chemical composition ($CaCO_3$), but aragonite is less stable and dissolves when buried in a sediment or recrystallizes to calcite. This is an example of diagenetic change. Nearly all mollusc shells (clams, snails) are composed of aragonite.

   Does this rock illustrate an example of recrystallization or selective dissolution?

2. Use a binocular microscope to determine the composition of carbonate grains in loose sediment samples. Use the sediment types of **bioclasts, abraded particles, ooids, and intraclasts**. Name the carbonate rock that would be produced by this sediment. Make an estimate of sorting of the sediment. Use the flowchart (Figure 2) to help identify sediments and rocks.

	Sediment Name	**Major grain**	**Secondary grain (if present)**	**Grain size**	**Sorting**
Sediment					
111					
112					
113					
114					
115					

3. Determine the rock type (name) of each sample, using the carbonate rock classification. List the main grain types and matrix present in each rock. Use one of the carbonate rock names listed above.

	Rock Name	Grain types	Matrix
37			
38			
39			
40			
50			
51			
58			
116			
117			
118			

4. Many sediments contain both carbonate and siliciclastic components and may not fit well in either classification. The boundary between carbonate and siliciclastic is set at 50% carbonate content, but transitional sediment types might be better named in the other classification if it indicates the depositional process producing the sediment.

   Name these two sedimentary rocks using the carbonate **and** the siliciclastic classifications. Decide which is a better name for the sediment and explain why.

   A. (33)

   B. (119)

5. Examine the specimens containing oncoids (oncoliths) (57) and algal balls (120). What character do oncoids (oncoliths) have that is different from ooids?

Sediments in Depositional Environments

6. Depositional Environment Review

Apply the information on sediments, siliciclastic, and carbonate, to major depositional environments that will be worked on in the next lab. Review information covered in Physical Geology to complete the table below and **bring the completed table to lab next week**. An example of sediment transport process is wave action on a beach; channel flow in a river. Each of these environments has a transport process or lithology that is distinctive and characteristic. Use names provided below.

	Typical sediment type (list one)	Major sediment transport process	Lithology or sedimentary process distinctive to environment
Reef			
Meandering river			
Swamp			
Shorezone/Beach			
Open continental shelf			
Deep ocean			

For sediment type, choose one of the following: **chalk, conglomerate, limestone, peat, sandstone, shale**.

For transport process, choose one of the following: **bottom currents, channel flow, glacier flow, settling from suspension, waves**.

For lithology or process, identify a rock type or transport process most distinctive for that environment.

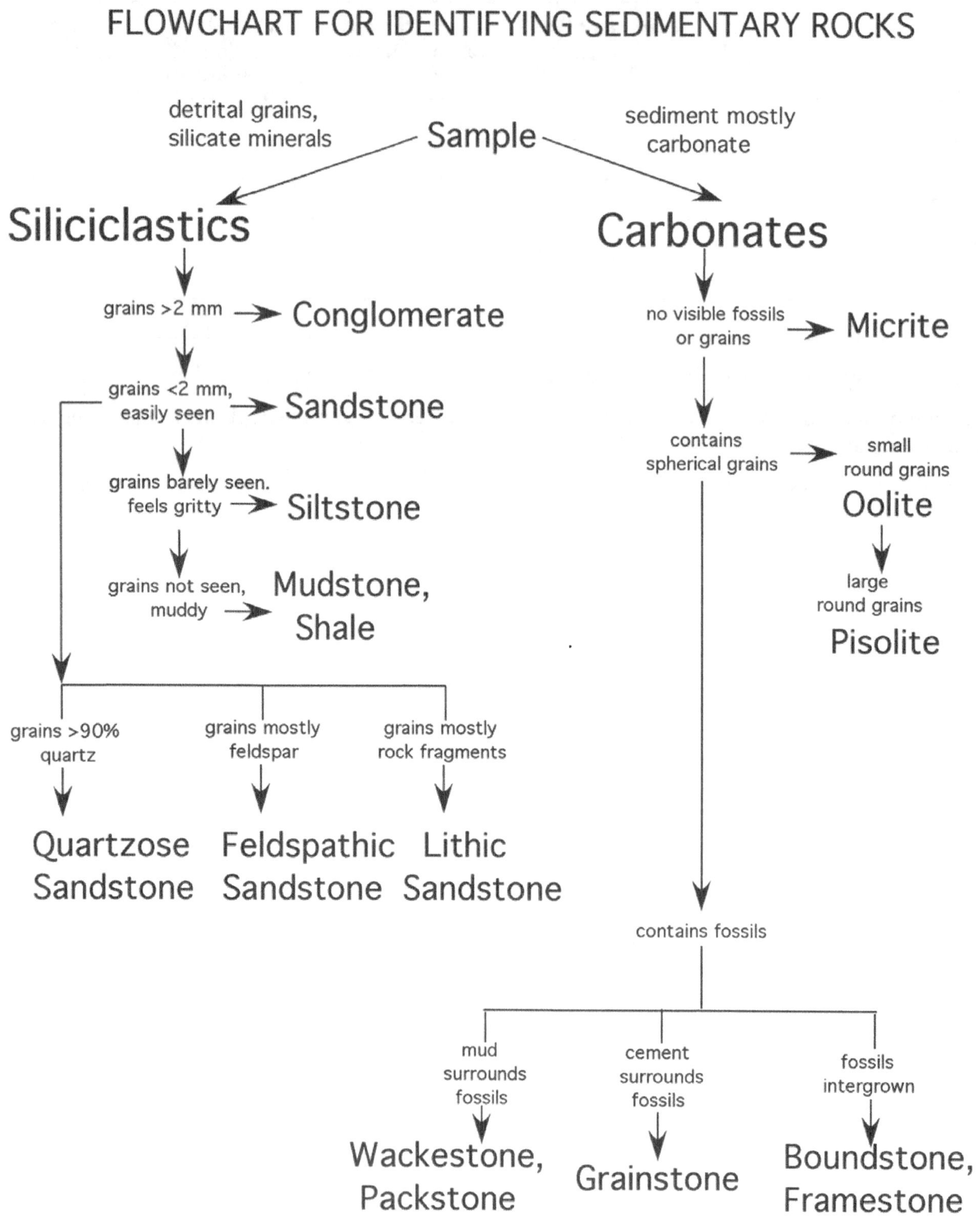

Figure 2 Flowchart for identifying sediments and sedimentary rocks.

Sedimentary Depositional Environments

Lab 4

Sedimentary Structures

Unidirectional flow sedimentary structures (Figure 1) are produced during times of decreasing energy level of a current or other transport process. Most sediments show fine scale layering—**lamination**—resulting from small variations in energy as grains continually fall out of suspension of the water/wind currents. Particles moving along the bottom at low velocities move as a migrating dune or ripple with an inclined front surface producing **cross bedding** with laminae inclined at 30°–35°, the maximum angle of stability for loose grains. Sediments showing no noticeable lamination are described as **massive**.

Bidirectional flow (alternating flow) is typical of tidal currents and produces **herringbone** and **flaser/lenticular bedding** (Figure 1). **Oscillatory motion** of waves in standing water produces ripples or hummocky cross -stratification (Figure 2).

Compound sedimentary structures are combinations of simple structures produced by changing depositional process. A good example is produced when turbidity currents (masses of sediment-carrying water flowing on the seafloor into deep basins) slow down and deposit sediment. Grain size decreases and bedding changes until all that is present is the normal fine mud of a deep-water basin.

Scour structures (Figure 3) are produced when previously deposited sediment is disturbed by the water (or a moving mass of sediment) and removes some sediment or reshapes the sediment surface, resulting in **scour or winnowing**. This indicates a short-term increase in energy conditions that produces some erosion, such as flute marks. Flute marks are long grooves made by dragging an object across a sediment surface.

Disturbance sedimentary structures (Figure 4) include **deformational structures** (convolute bedding, load structures, ball and pillow). Load structures and ball and pillow structure are the result of density differences in the sediment, when heavier sediments (sands) sink down into lighter sediments (mud) or lighter sediment pushes upward. **Convolute bedding** develops when fluids rise through overlying sediments, distorting them, or when sediment masses begin to flow (slumping). **Bioturbation** (disturbance by organisms) also produces

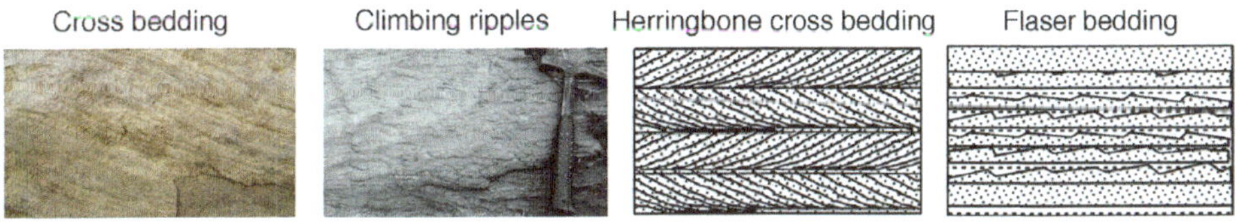

Figure 1 Unidirectional and bidirectional sedimentary structures.

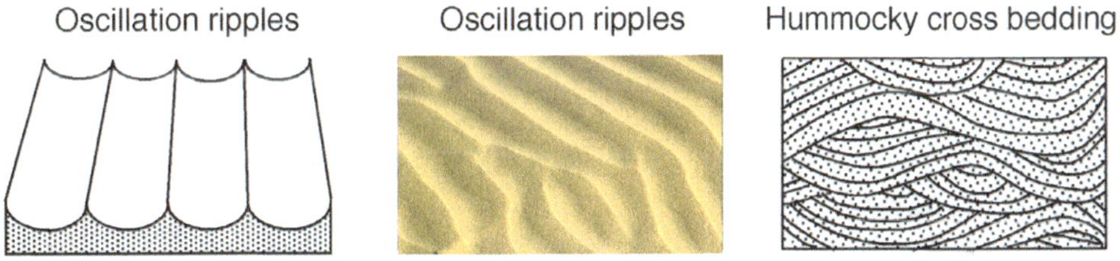

Figure 2 Oscillatory motion structures.

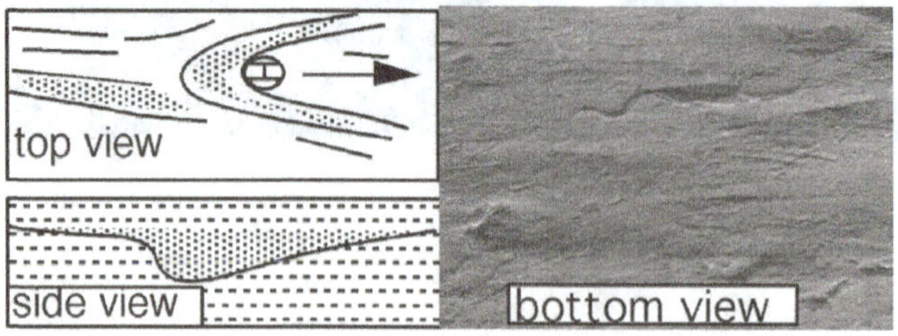

Figure 3 Scour structures.

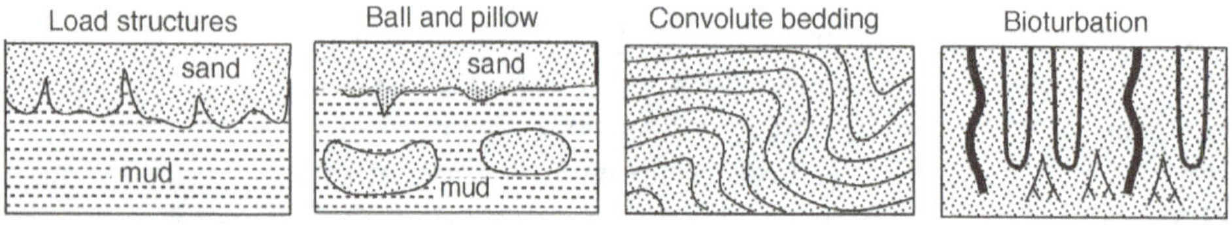

Figure 4 Secondary sedimentary structures.

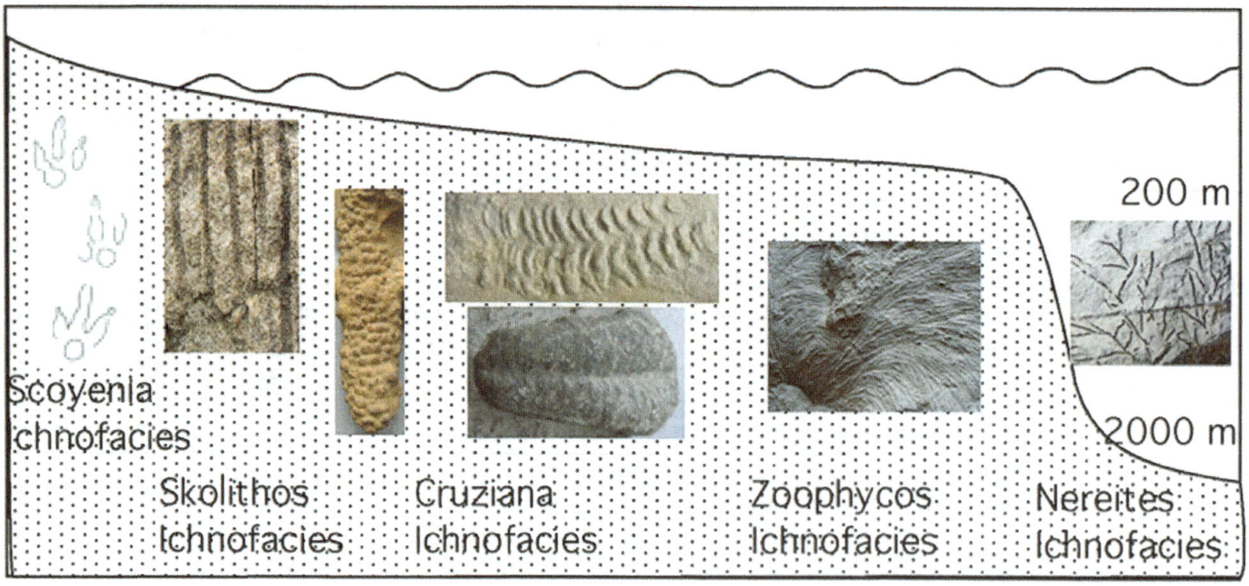

Figure 5 Trace fossil zonation in depositional environments.

secondary structures (tracks, burrows, root traces—all are trace fossils), by animals living in sediments or plants growing on soil.

Subaerial (exposure) disturbance structures are produced in sediments exposed to subaerial conditions, where they usually develop a variety of structures specific to conditions of drying. The most obvious indicator is **mudcracks**, which develop as clayey sediments shrink during water loss, opening cracks that we see as mudcracks in the rock. Other indicators include **rain drop imprints**, occurrence of **rooting structures** (root traces), and development of **fossil soils**. Fossil soils (called paleosols), are indicators of long exposure, associated with lack of sediment deposition or erosion on that surface.

Trace fossils (Figure 5) are **bioturbation disturbance structures made by organisms**. They provide evidence of organisms present and are indicators of the amount of oxygen in the sediment. Some trace fossils can often be used as a rough indicator of water depth. A greater use of trace fossils is to interpret the life activities of the organisms. **Burrows** are the most common type of trace fossil. Other types include tracks and trails, feeding, shelter, boring, and rooting traces.

Deep vertical burrowing is common in shallow water, sand substrate environments; shallow horizontal traces are normal in deep-water mud-bottom environments.

Exercise Work

1. For the specimens present in the lab set, identify the type of sedimentary structure and make inferences about the depositional processes present in the environment at the time of deposition or the type of disturbance affecting the sediment shortly after deposition. Refer to the drawings (Figures 2–5) illustrating a variety of sedimentary structures for help in identifying the structures. For Depositional Process, use terms: **unidirectional, bidirectional, oscillatory, scour, disturbance**, **subaerial disturbance** or **bioturbation** for the current or process.

	Sedimentary Structure	**Depositional Process (type of current, etc.)**
#64		
#66		
#70		
#75		
#76		
#78		
#98		
#305		
#343		
#344		
#366		

2. **A.** This turbidite specimen (#422) consists of a compound sedimentary structure that is the hallmark of a turbidity current deposition. Refer to Figure 7: what is the depositional environment with turbidites?

   B. The dark bands at the top and bottom of the samples (#422) are quiet water deposits, enclosing (sandwiching) the deposits of higher energy turbidity currents. Deposition from a turbidity current produces sediment with a regular change in grain size, named a Bouma sequence. What is the expected change in grain size: upward-fining grain size or upward-coarsening grain size in a turbidite unit?

   C. What causes a turbibidy current to deposit sediment of changing grain size through time?

3. **A.** What depositional environment is indicated by the sedimentary structure in this sample (#33)?

   B. What process of sediment transport could deposit this type of sediment?

4. **A.** What depositional environment is indicated by the lithology and sedimentary structure of this sample (#30)?

   B. Describe the conditions of sediment accumulation in this environment (make an inference)?

   C. What conditions were present to allow for preservation of organic matter?

5. **A.** What depositional environment is indicated by the compound structure of this specimen (#51)?

   B. Make an estimate (in percentage) of the skeleton versus matrix content of this specimen (#51).

6. **A.** What depositional environment is indicated by the sedimentary structure of this specimen (#64)?

   B. What high energy event is responsible for producing this type of deposit?

7. **A.** What type of bedding occurs in the sand of this sample (#63)?

   B. This has the form of a ripple; is this produced by unidirectional or oscillatory flow?

   C. The specimen is an example of a compound structure. Among the depositional environments listed in Figure 7, what depositional environment(s) is indicated by the bedding and sedimentary structures (list all that could apply)? Which one is most probable? Consider that this could be part of a sediment suite and not a complete suite.

8. Simple cross bedding is produced by unidirectional water flow whereas bidirectional cross bedding is produced by alternating directions of water flow. On an environmental transect from alluvial fan to open shelf, see Figure 6, in what environment transition will water movement change from unidirectional flow to back and forth movement and change from depositing simple cross beds to bidirectional cross beds?

Depositional Environments

There is a continuum of environments present on the Earth surface and they can be divided into fairly distinctive depositional environments (Figure 6). Each has characteristic depositional processes producing distinctive suites of sediments and sedimentary structures (Figures 1–5) and may contain fossils. Environments vary along elevation/bathymetric gradients like the proximal versus distal nature of deposits in alluvial fans, river systems, deltas, open shelves, and submarine fans. Other environments are present along rainfall gradients (high rainfall to desert) and temperature climate gradients (poles to the equator).

Depositional environments are recognized by suites of characteristic 1) sediment lithologies and 2) sedimentary structures present in the sediments. They also can be determined from the 3) fossil content of the sediments.

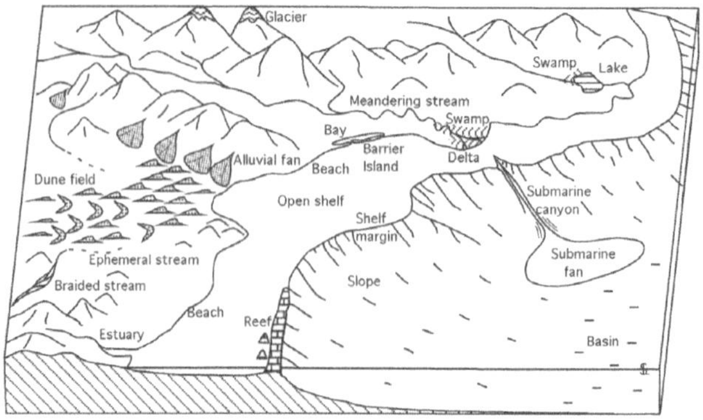

Figure 6 Overview of sedimentary depositional environments

Depositional Environment Suites of Sediments and Sedimentary Structures

Major depositional environments are identified by lithologies and sedimentary structures. Some have nearly uniform sediments (Alluvial Fan; Swamps; Beach), but most contain groups of lithologies and compound structures produced by changing transport processes and falling energy level (Meandering River; Storm Shelf; Submarine Fan). Where present, these are reliable indicators of depositional environment (Figure 7).

Alluvial Fan: Alluvial fans contain poorly sorted sediments. **Depositional processes: sheet wash and mass flows.** On steep slopes, heavy rainfall and/or landslides weaken soil and surface rock, produces mass flows with very poor sorting and a wide range of grain size. Mass flows deposit sediment when friction causes the flow to stop moving. When mass flows are present, it is an indication of alluvial fan environment.

Meandering River: Meandering rivers continually move gravel, sand, and fine sediment with gravel moving during occasional floods. **Depositional process: channel flow.** This produces an **upward-fining set of sediments with gravel at the base and mud at the top.** As the river channel migrates and changes locations it produces wide sheets of upward-fining sediments.

Swamp: Swamps contain fine-grained muds with high organic content, including peat. **Depositional process: still, quiet water.** Sediments lack current indicators.

Shorezone/Beach: Shorezone deposits are well sorted sands that may contain shell-rich layers or some pebbles. **Depositional process: regular wave action of water** washing back and forth.

Ocean Shelf (storm): Ocean shelf deposits consist of mud and silt deposited by settling from suspension, either as mud or from plankton in the water or infrequent storms. **Depositional process: Settling or brief storm currents.** Storm deposits are common in most shelf deposits as upward-fining graded sediment enclosed in mud deposits. The characteristic storm deposit is hummocky cross-bedded sand. In warm oceans, shelf deposits can contain abundant carbonate shell and even reefs.

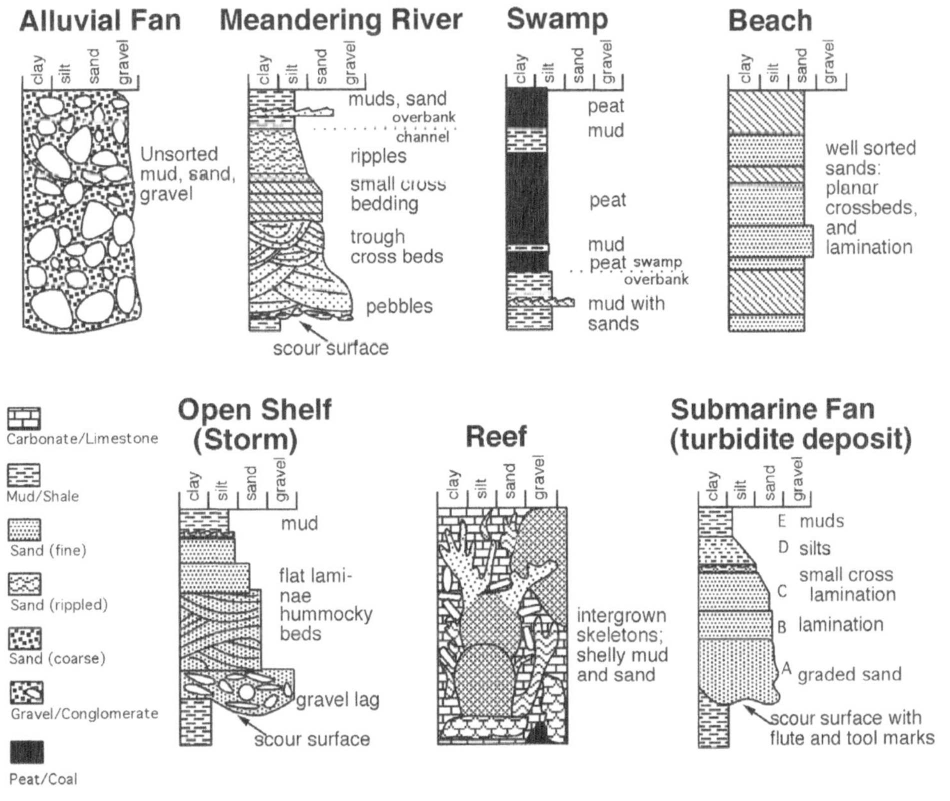

Figure 7 Typical sediment deposits of sedimentary depositional environments.

Reef: Reefs are characterized by the presence of intergrown calcareous skeletons surrounded by loose sediment. **Depositional process: growth of organisms.** Most sediment in a reef is loose formed by the breakdown of skeletons.

Submarine (turbidite) Fan: Submarine fans grow by sediment moving down the continental slope as a turbidity current. **Depositional process: density bottom current.** As the flow loses energy, sediment of smaller grain size is deposited, producing a graded upward-fining turbidite deposit called a Bouma sequence. Channel deposits and overbank deposits are present in a fan.

Exercise Work

9. Using the diagrams shown in Figure 8: Identify the basic depositional environment (alluvial fan, braided river, meandering river, swamp, beach, storm shelf, reef, and submarine fan) of each of the sediment suites shown below, **filling in the name on the line above the diagram.** The braided river environment, not discussed before, is labeled here. Each of these represents a distinct sedimentary depositional environment (a facies) on a depositional surface.

10. Next, plot these on the accompanying map sheet (Figure 9), using sample numbers. Sample locations with the same sediment suite are a geographic area covered by a depositional environment. **Draw boundary lines between these geographic districts and name the depositional environment present in each area. Adjacent districts share a common boundary with no gaps.** All eight depositional environments are present on the map; **use only the eight named environments.**

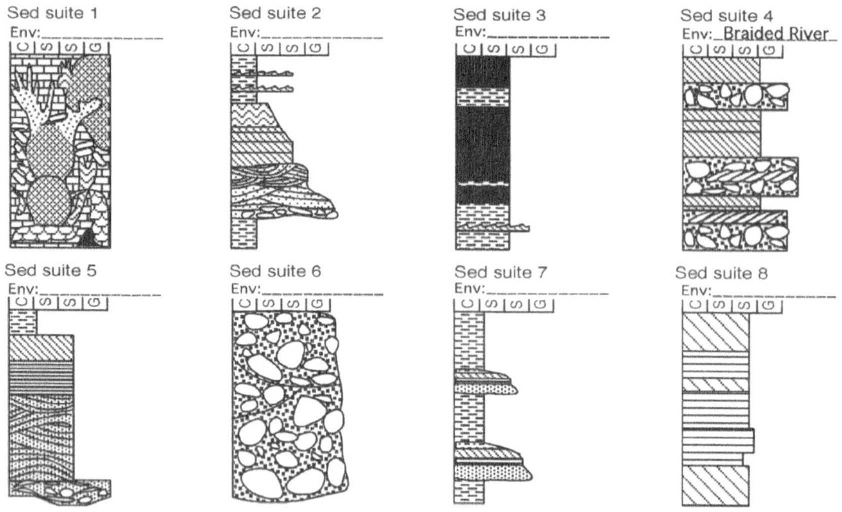

Figure 8 Typical sediment deposits of sedimentary depositional environments. The letters C, S, S, and G in each diagram refer to the sediment grain size ranges of: clay, silt, sand, and gravel.

11. Using a blank piece of paper, draw a profile of the probable topographic relief along a line extending from A to B on the map, which crosses the combined land and ocean floor surface. Draw a horizontal line across the center of the page to mark sea level and draw the profile of mountains, shelf, and basin relative to this datum.

12. **A.** The deposits plotted on Figure 9 show the distribution of depositional environments (**facies**) and lithologies present in a sedimetary basin. When the sediments are buried by younger strata, assess the potential for production of fresh water or for petroleum extraction from these facies. Calculate the area of each facies by estimating the average width and the average length of each, then multiply.

#1 Northwest	#1 East	#2	#3	#4	#5	#6	#7	#8

B. Consider the porosity of sediment in each facies. Which environments contain sediment with the porosity and permeability to be aquifers for water or reservoirs for petroleum?

13. What depositional environment (facies) would contain sedimentary structures like that of samples:

A. #64

B. #305

C. #422

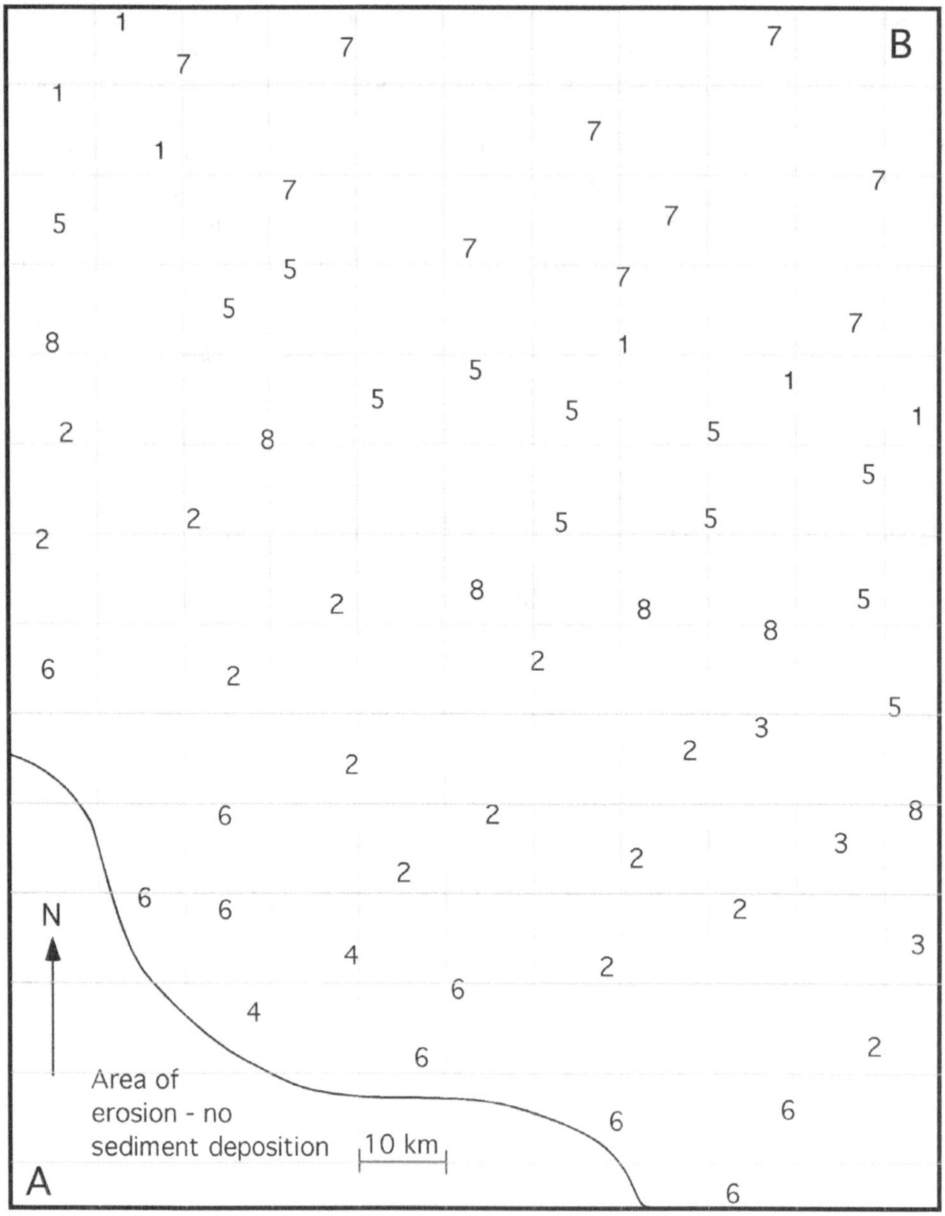

Figure 9 Base map for plotting depositional environments.

Provenance: Sediment Source

Lab 5

Introduction

Determining the source of sediment is known as the study of **provenance**. Mineral composition of sediments is controlled by the rocks that are eroded, so identifing the mineralogy of the sediment helps identify source rocks and the geology of the source terrain. Because chemical weathering preferentially destroys less stable minerals, most sediment is composed of stable minerals. Despite these limitations, provenance study is an important tool in determining the **paleogeography and geologic history of an area**. In addition, information about ancient **climate** of land areas may be determined from the degree of mineral alteration.

Chemical weathering removes unstable minerals and concentrates more stable minerals, a process called **mineralogic maturity**. Long-term transport will usually modify the sediment particles and produce more rounded and spherical grains, a process called **textural maturity**.

Easily weathered minerals	Minerals resistant to weathering
Olivine, pyroxenes, amphiboles, feldspar	Quartz, mica, zircon, clays, Fe, and Al oxides

Exercise Work

1. Sediment samples A, B, C, and D have different maturities. Using the weathering index shown above, arrange the samples in a series from least mineralogic mature to most mature. During chemical weathering on the Earth's surface, there is a concentration of quartz and clays.

   Criteria for mineralogical maturity: If you can see some olivine, amphibole, or pyroxene (usually seen as black grains), or pieces of basalt rock, or grains of carbonate (eroded from limestones) in a sediment, call it **immature**. If the sediment has no olivine, amphibole, or pyroxene but has a few percent or more of the feldspars and micas, call it **submature**. If the sediment is nearly all quartz or resistant oxides, call it **mature**. Review the mineral groups presented in Laboratory 2.

Mineralogically immature						Mineralogically highly mature

2. Using the charts of sorting and roundness shown in Lab 3, arrange sediment samples A, B, C, and D in a series from least texturally mature sediment to most mature. Grains in a sediment usually vary in roundness, so determine an average. If roundness and sphericity suggest different levels of maturity, use roundness as the best indicator. Ignore platy minerals like mica grains.

Criteria for textural maturity: If grains are angular and some have low sphericity, call the sample **immature.**

If the grains are partly rounded and subspherical, call the sample **submature.**

If the grains are well rounded and spherical, call the sample **mature.**

Texturally immature						Texturallyhighly mature

3. Sediment samples A, B, C, and D have been weathered from four different parent rock types: Granite, Basalt, Quartzose sandstone, and Lithic sandstone (i.e., sandstone with many multimineral rock fragments). Examine each sediment sample and match the sediment sample with the parent rock type.

Sediment Sample	Parent Rock Type
	Quartzose sandstone
	Lithic sandstone
	Granite
	Basalt

Interpretation of Source Terrain

Mineral composition of sediments is controlled by the rock from which sediments are eroded, so sediment mineral composition is used to interpret the geology of source terrains. There are three general types of source terrains that produce the majority of sediments—**Sedimentary, Plutonic-Metamorphic,** and **Volcanic**—but some less common types produce very distinctive and easily recognized sediments such as **Subduction** zone rocks (Table 1).

Sedimentary rocks are formed under Earth surface conditions and sediments derived from them contain stable minerals (quartz, clays, etc.) or have easily recognized sedimentary textures.

Plutonic-metamorphic rocks are of deep subsurface origin in conditions of high temperature and pressure, and commonly have large crystal size. The dominant minerals tend to be unstable in surface weathering conditions, producing a sediment of limited compositional range but containing some distinctive minerals.

Volcanic rocks are usually fine grained because they are extrusive or intruded near the Earth surface, so they typically have a glassy texture or fine-grained matrix and a distinctive appearance. This provides a means of distinguishing most volcanic rock from plutonic-metamorphic rock. For this work, the category includes volcaniclastics composed of ash and cinder deposits.

Subduction: Among the less common types of source terrain producing distinctive sediment is that associated with subduction zones; a subset of the igneous-metamorphic category. It contains very distinctive rock types and is an association commonly found within tectonically active continental margins. Subduction rocks contain minerals formed in conditions of higher pressure combined with lower temperature than the normal Plutonic-Metamorphic type. Unusual rocks like glaucophane schist (called blueschist) and eclogite are typical and often associated with mantle peridotite and pieces of ocean floor crust that have been altered to serpentine.

Table 1 Rocks and Minerals of Major Source Terrains

Sedimentary	Plutonic–Metamorphic	Volcanic	Subduction
sandstone conglomerate shale limestone coal dolostone chert	gneiss schist phyllite slate quartzite granite diorite vein quartz	basalt andesite gabbro rhyolite obsidian volcaniclastics olivine	eclogite blueschist (glaucophane) peridotite serpentine jadeite garnet

4. Examine the sedimentary rocks listed below and interpret the provenance (source terrain) of the sediments. Identify the rock type of large clasts in the rock and the mineralogy of small grains to form a judgement about source terrain. **Use the categories of: sedimentary; plutonic-metamorphic; volcanic.** If the clasts suggest a mixture of source terrains, identify the terrain which provided the majority of the sediment. Record the evidence for each interpretation.

	Type of Source Terrain	Evidence
#28		
#52		
#54		
#73		

5. Large rivers will usually carry sediment from multiple source terrains, with tributaries delivering sediment of different composition. The map below (Figure 1) shows the Cordillera River drainage basin. It grew in size as tectonic uplift and deformation led the river to capture streams draining source terrains with distinctive rock assemblages.

The **Sedimentary** terrain contains shallow marine sedimentay rocks.

The **Plutonic–metamorphic** terrain contains common granites, mica schists, and slate rocks.

The **Volcanic** belt contains basalt and rhyolite rocks and volcanic ash.

The **Subduction** terrain contains peridotite, amphibolite, blueschist and eclogite rocks.

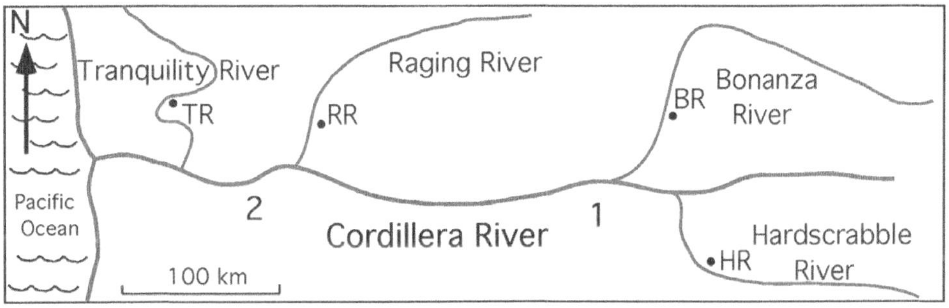

Figure 1 The Cordillera River system with tributary rivers.

Use river samples Tranquility River (TR), Raging River (RR), Bonanza River (BR), and Hardscrabble River (HR) to identify the type of source terrain present in each of the tributary rivers of the river system.

Sample TR: Clasts of sandstone, chert, and limestone. Sands contain grains of chert and quartz.

Sample RR: Clasts of granite, gneiss, mica schist, slate, and vein quartz. sands contain tourmaline and mica flakes in addition to feldspar and quartz.

Sample BR. Clasts of basalt, rhyolite, and obsidian. Sands contain grains of amphibole, feldspar, and rare olivine.

Sample HR: Clasts of amphibolite, blueschist, and eclogite. Sands contain grains of pyroxene, amphibole, and garnet in addition to quartz.

TRIBUTARY RIVER	TYPE OF SOURCE ROCK TERRAIN
Tranquility River	
Raging River	
Hardscrabble River	
Bonanza River	

6. Sediment in the main river channel is a mixture of sediment from the tributaries. Each tributary adds a distinctive suite of sediment and main channel sediment changes and becomes more diverse from the headwaters to the mouth of the river. List the composition of the gravel clasts and mineral grains found in sediments of the river channel at stations 1 and 2.

Station 1	Clasts: Mineral grains:
Station 2	Clasts: Mineral grains:

Detrital Zircons and Tectonic History

Zircon is the most useful mineral in sediments for determining provenance and interpreting tectonic history. Zircon crystallizes in most igneous rock and persists in sediments because it is very durable and resistant to chemical weathering. It contains uranium and single crystals can be datable by U-Pb decay. Most zircon occurs as small sand-sized crystals, so is present in most sand deposits. Dating each zircon grain in a sample gives an age profile that is compared to ages of known source areas. For example, an abundance of zircons with ages of 500–300 Ma corresponds with the Alleghenian Orogeny and indicates an origin in the Appalachian Mountains; an age of 1,250–950 Ma corresponds with the Grenville Orogeny. Changes in zircon ages through geologic time reveals tectonic history and sediment transport.

An example of an age profile of detrial zircons in a sediment sample is shown in Figure 2.

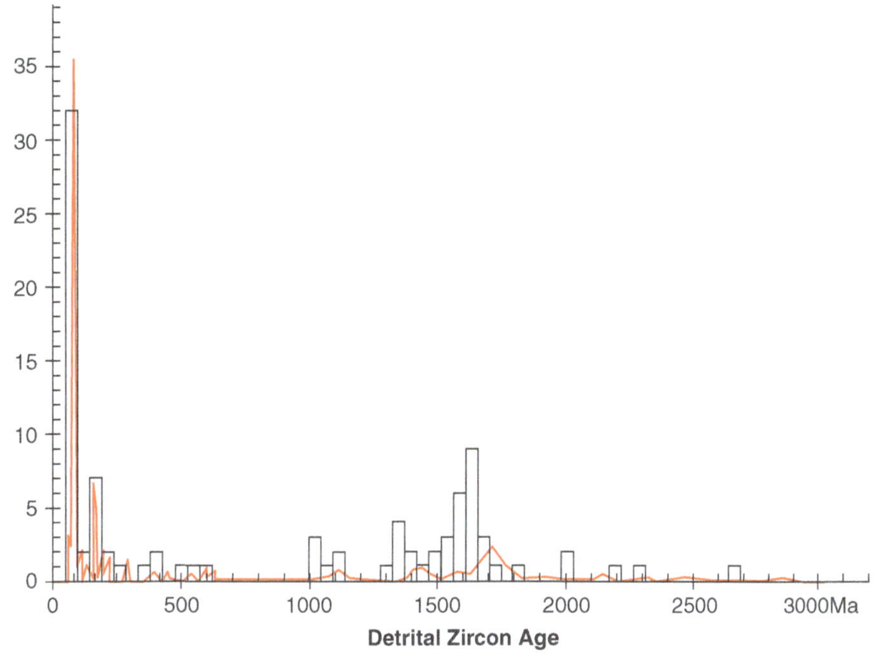

Data from A. Hutto thesis (2009).

Figure 2 Relative frequency of ages of detrital zircon grains in a sample from the Wilcox Group of east Texas.

Figure 3 Bedrock (crystalline igneous-metamorphic) geologic provinces of North America. Basement rocks in each province have a similar age produced by recrystallization during igneous-metamorphic deformation.

Table 2 Crystalline Bedrock Ages of Primary Geologic Provinces in Cental North America

Geologic Province	Age of crystalline bedrock in millions of years
Laramide	80–50 Ma
Cordilleran arc (subduction margin)	270–50 Ma
Allegheniann (Appalachians)	500–300 Ma
Grenville	1,250–950 Ma
Mid-Continent	1,500–1,300 Ma
Yavapai-Mazatzal	1,800–1,600 Ma
Trans-Hudson-Penokian	2,000–1,800 Ma
other Canadian Shield	>2,500 Ma

Exercise Work

7. A. Using Figure 2 above and the age data of crystalline basement rocks in North American geologic provinces listed in Table 2, determine the source of the three dominant modes of detrital zircons shown in the Wilcox Group.

   B. Use that data to detemine the direction of sediment transport to Texas at the time of Wilcox Group deposition.

8. A. Use Figure 4 below and age data of basement rocks in Table 2 to determine the source of sediments present in the Brushy Mountain Formation of the Permian Basin in west Texas.

   B. Detemine the direction of sediment transport to Texas at the time the Delaware Mt. Group was deposited.

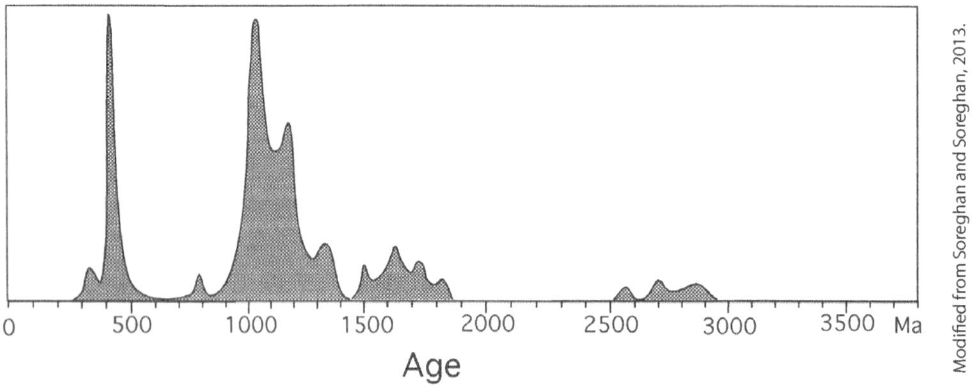

Figure 4 Relative frequency of ages of detrital zircon grains in samples from the Brushy Canyon Formation of west Texas. Age groups correspond with major tectonic events..

Tectonic Unroofing History

In stratigraphic sections deposited within a basin near a mountain range, a record of change is preserved in the form of sediment and rock clasts shed from the mountains. Uplift of a mountain range exposes and erodes rocks from the top-down, so the stratigraphic section contains an inverted succession of rock types. A series of erosional conditions is shown in Figure 5, where A is the first time slice and C is the last time.

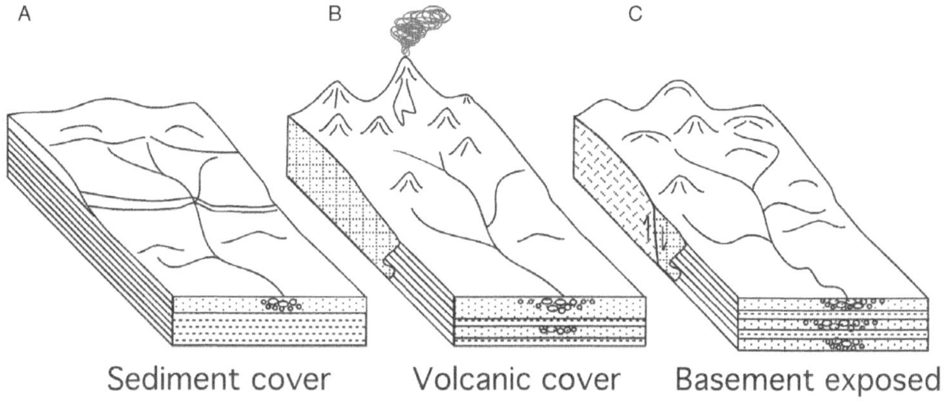

Figure 5 Time series showing first erosion of sediment strata, later erosion of volcanics, and last erosion of uplifted plutonic–metamorphic rocks.

9. Identify the rock assemblages of Figure 6 in terms of **Plutonic-metamorphic; sedimentary, or volcanic** and determine the geologic history of mountain range development. From these determinations, write a brief geologic history of the development of the mountain range and the rock types that were exposed through time based on the concept that during uplift exposure, sediment eroded from the highest rocks are deposited first and sediments derived from lower levels are deposited later. Outline a geologic history that explains all the geologic events indicated.

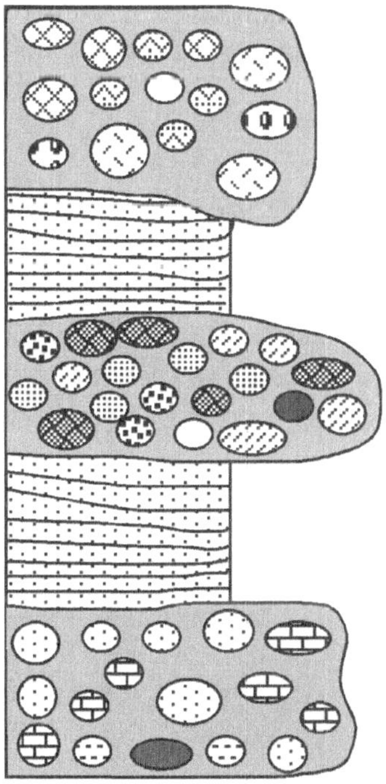

This conglomerate contains pebble and cobble clasts of granite, gneiss, diorite, and mica schist.

Source rock assemblage? ____________________

This conglomerate contains pebble and cobble clasts of andesite, rhyolite, lithic sandstone, and vein quartz.

Source rock assemblage? ____________________

This conglomerate contains pebble and cobble clasts of quartzose sandstone, limestone, and chert.

Source rock assemblage? ____________________

Figure 6 Stratigraphic section containing sediments eroded from adjacent landmass and deposited in a basin.

Correlation and Age Relations of Strata

Lab 6

Introduction

The resolution of many questions in geology is dependent on knowledge of geologic age. This is done in two ways: calculation of numerical age and determination of relative age. Many times it is not possible to determine the numerical age of rock units or events, but it is possible to determine the *sequence* of events that occurred and the **relative age** of the rocks by comparison with other rock units.

Relative Age Dating

Relative age dating, determining older versus younger, is developed from the Principle of Superposition. This relationship enabled early workers to determine age and produce the first geologic maps. By the early 1800s, William Smith and Georges Cuvier developed procedures for using fossils to determine relative age of strata in widely separated regions.

The *Principle of Superposition* states that in any undisturbed sequence of extrusive or sedimentary rocks, the oldest rocks are deposited first and the younger rocks are deposited above them. Therefore, older rocks occur at the bottom (Figure 1).

Relative ages are also determined with the concept of Cross-cutting Relationships, which states that any rock unit or structure that cuts across other rocks or structures must be younger than the rocks or structures that it cuts (Figure 2).

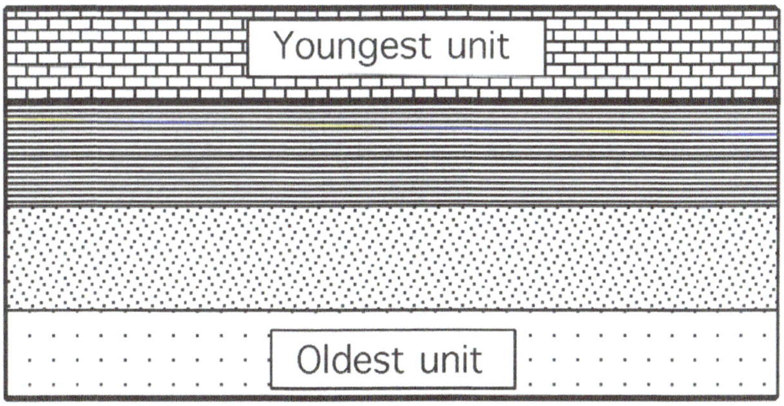

Figure 1 Principle of Superposition.

Unconformities

Another relationship that is important in determining relative ages is the **unconformity**, surfaces that represent periods of nondeposition and/or erosion resulting in loss of some portion of the geologic record. Although there is no direct record of the event, the form of unconformity can suggest the geologic processes operating during that time. An unconformity often has time significance within a region and may be useful for correlation. There are three types of unconformities. A **disconformity** is an unconformity in which a major erosional surface separates parallel layers of sedimentary rock. An **angular unconformity** is an unconformity in which the layers below the erosional surface are *not* parallel to the layers above it, because the older layers were tilted and eroded before the upper layers were deposited. A **nonconformity** is an unconformity that forms when older igneous or metamorphic rocks are eroded and exposed and younger sediments are deposited above.

The relationship between these types of unconformity can be seen in Figure 4, which shows the changing character of an unconformity across a basin and nearby highlands, providing information on the geologic history of the area. Exposure to erosion following a drop in sea level produced a basin-wide unconformity, which is a

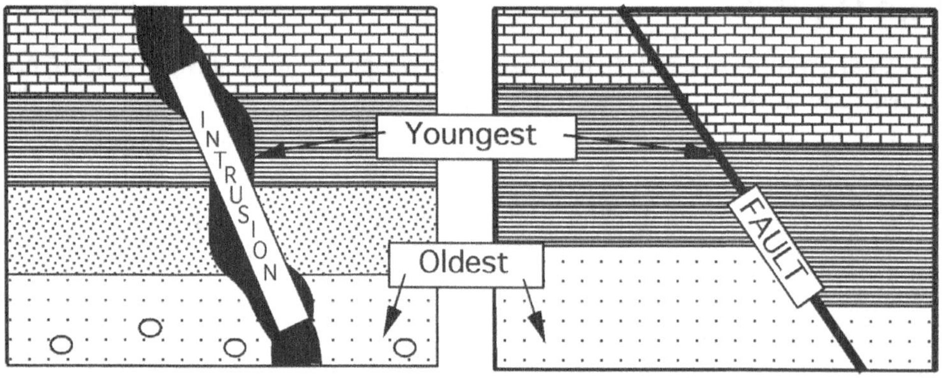

Figure 2 Cross-cutting relations of rock units.

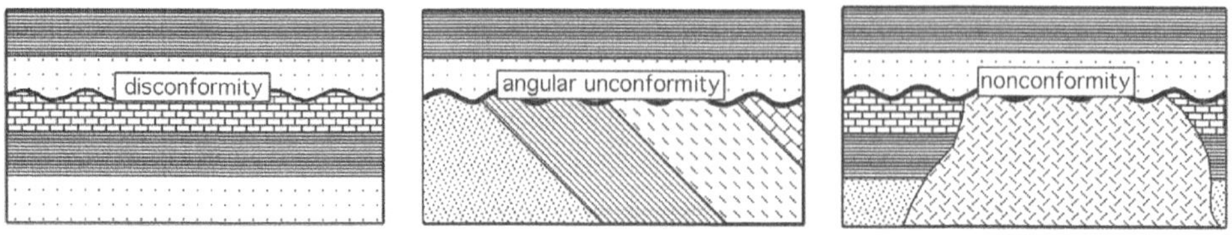

Figure 3 Disconformity Angular Unconformity Nonconformity

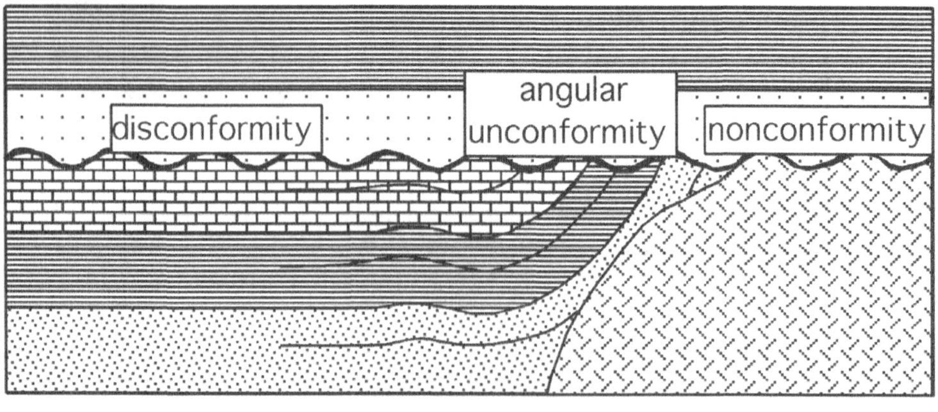

Figure 4 Unconformity variation across a basin.

disconformity in the center of the basin. On the basin margin, an igneous intrusion deformed the sedimentary rocks, tilting them so that erosion produced an angular unconformity. Where erosion uncovered the top of the igneous intrusion, a nonconformity was produced.

Ordering of Events

Relative dating is often the basis for determining even the most complicated geologic histories. Study the cross section below (Figure 5) as an example of the procedure used for unraveling geologic history.

No overturned beds are present, so we can infer (from the Principle of Superposition) that **bed U** is the youngest and **bed C** is the oldest. Two breaks in the record, both being angular unconformities, are shown. (Remember that an angular unconformity is formed when beds are tilted and eroded, followed by deposition on the eroded surface.) Relationship between **beds F** and **M** or **beds O** and **T**, which lay on opposite sides of the unconformities, shows which bed is older and which is younger. For our work, we will assume that an erosion surface forms as a nearly horizontal surface. The geologic history shown in the cross section (Figure 5) is as follows (in time sequence from the oldest to the youngest):

1. Oldest event: successive deposition of **beds C, D, E,** and **F** (superposition)
2. Uplift of the area in the east results in these beds being tilted and exposed to erosion.
3. Erosion of beds C–F producing unconformity
4. Subsidence followed by successive deposition of beds M, N, and O
5. Renewed uplift, this time in the west, tilting both sets of beds and exposing them to erosion
6. Erosion of beds C–F and M–O producing unconformity
7. Subsidence followed by the successive deposition of beds **T** and **U** (last event).

From this work we are able to determine the relative dates of the rock units (C through U), times of erosion and non deposition in the area, and by identifying the type of unconformity, we can infer tectonic events affecting the area (uplift and tilting of rock units).

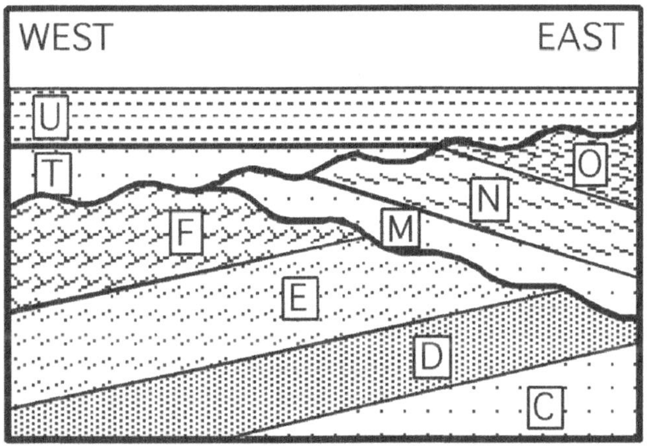

Figure 5 Geologic cross section.

Correlation

Stratigraphy is the branch of geology that deals with the study of layered rocks, or *strata*. The fundamental unit of rocks used in stratigraphy is the formation. A **formation** is a body of rock that has enough distinctive characteristics so that it may be mapped as a unit. The primary characteristic used to define a formation is the rock type, or **lithology**, and the boundary between formations is called a **contact**. The lithology of rocks contained in a formation may be somewhat variable, but these rocks are similar enough to be distinguished from those of adjoining formations.

The formations present in an area can be shown in a **stratigraphic section** (Figure 6) (also called a geologic column), or vertical sequence of rock units, that show the position and relative age of the formations. By comparing stratigraphic sections from different areas, the lateral extent and thickness variations of formations can be determined in a process of **correlation**, the determination of equivalency of rock units. Correlation of rock units can be done on the basis of rock type (lithostratigraphic units), age (chronostratigraphic units), or fossil content (biostratigraphic units). Correlation of stratigraphic sections is important for the study of depositional environment, tectonic setting, and paleoclimate.

The basis for the study of strata and stratigraphic correlation is the **Principle of Superposition** and its corollaries, Original Horizontality and Lateral Continuity.

This principle states that in any undisturbed sequence of sedimentary or extrusive rocks, the oldest rocks are deposited first and the younger rocks are deposited above them. Therefore, older rocks will be at the bottom and younger rocks at the top. A corollary to the law of superposition, the Corollary of Original Horizontality, states that sediments are deposited in horizontal or nearly horizontal layers, or as drapes over pre-existing surfaces.

Another corollary, the Corollary of Lateral Continuity, states that strata originally extended in all directions until they thinned to zero at the edge of the original area of deposition. The concept of lateral continuity is the basis for **correlation of strata**, which may be seen only as isolated, widely separated remnants of an originally continuous unit (Figure 7).

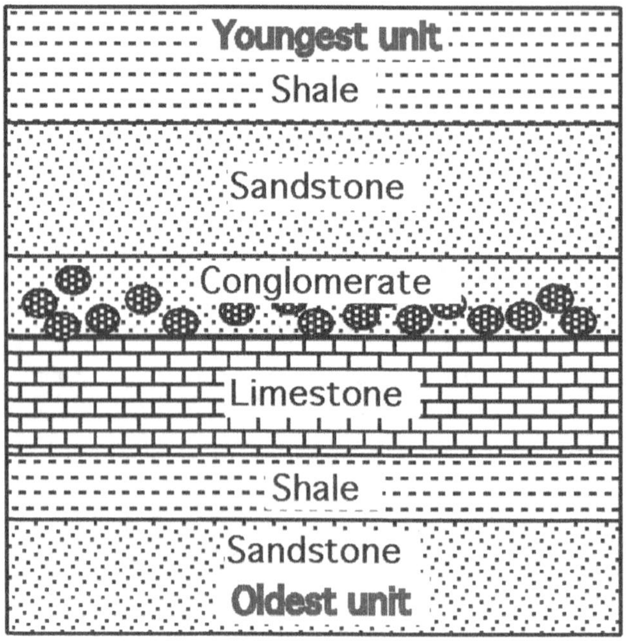

Figure 6 Stratigraphic section, showing age relations of stratigraphic units and illustrating the Principle of Superposition.

Lithologic Correlation

Working with strata involves a bed-by-bed description of their physical characters: the lithology, grain composition, bedding, sedimentary structures, fossil content, and the nature of boundaries between beds. The general character of the beds in a stratigraphic section is determined either from descriptions of surface outcrops or drillhole records and core. Stratigraphic sections are compared and matched to recognize originally continuous beds and determine their lateral and vertical extent and internal variation, a process called **lithologic correlation** (or **lithostratigraphy**). The process of lithologic correlation is made easier by the presence of marker beds, event horizons, or distinctive groups of beds. Stratigraphic units commonly vary in thickness laterally due to different rates of sedimentation or lateral change in depositional conditions leading to pinchout. Beds and lithologic units may also be laterally continuous but not of the same age everywhere (time transgressive strata).

Age Correlation and Sedimentary Facies

Correlation of deposits to show age equivalency is called **chronologic correlation** (or **chronostratigraphy**). Time horizons usually cut across lithologic boundaries because the lithology of deposited sediment is controlled by the geographic limits of a depositional environment, which can change position through time. Lithologic units show the extent of depositional environments and their shifts in position through time, while chronologic units show the sediments deposited during an instant or interval of time.

Many types of sediments are being deposited in depositional environments at the same time period. On a transect away from a shoreline, the high-energy beach zone contains sand sediment, proximal offshore environments have silt sediment, and further offshore environments have clay-rich sediment, corresponding to change from the high-energy end to the low-energy end of an energy gradient. The distal offshore shelf may also have a zone of carbonate sediment accumulation, produced by the accumulation of organism skeletons. These deposits are **facies** of each other—sediments that are lithologically different but time equivalent (Figure 8).

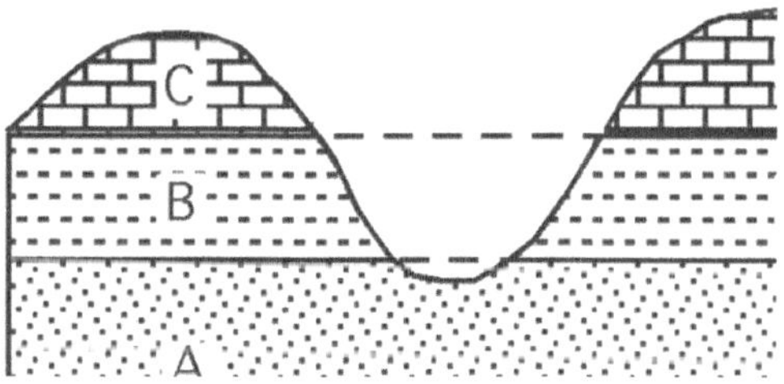

Figure 7 Lateral Continuity: reconstruction showing the relations of once continuous strata that have been eroded, leaving isolated remnants of the units.

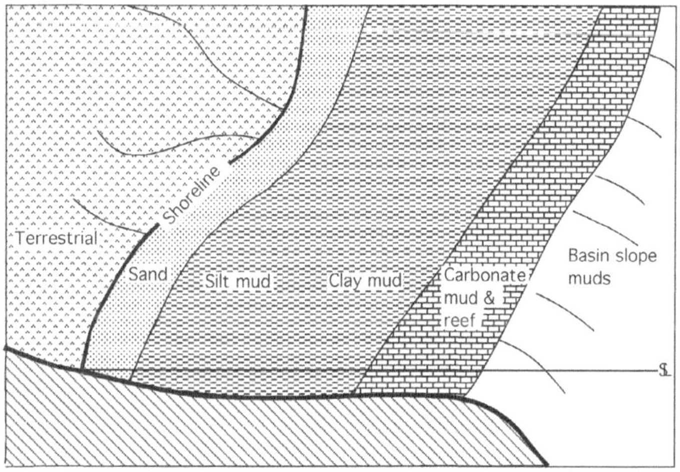

Figure 8 Depositional facies belts of sediments.

Transgressive–Regressive Cycles and Walther's Law

The normal consequence of sedimentation is an infilling of depositional environments, leading to shifts in position of these environments over time. As a basin is filled, the shoreline migrates toward the basin center and adjoining facies migrate in tandem, changing location along with the shoreline. This ends when the basin is completely filled.

The same type of change occurs in ocean basins during a rise or fall in sea level. If sea level rises, the shoreline changes position, water depths increase, and environments (and facies) change position and shift landward. This flooding by rising waters is a **transgression**. If sea level drops, the shoreline changes position, water depths decrease, and environments (and facies) shift seaward. This shoaling by lowering waters is called a **regression**. The vertical record of sediment deposited during a transgression and a regression are shown in Figure 9.

Changes in shoreline position occur continually, generating many episodes of landward migration of a shoreline followed by seaward migration of a shoreline in **transgressive–regressive cycles** (**T–R** cycles). This is seen locally as a period of rising sea level followed by falling sea level. A plot of time horizons extending across the facies tracts and the lithologic units shows that lithologic units are time transgressive. Figure 10 illustrates how environments of deposition migrate during transgression and regression, the positions of time lines in T–R cycles, and the lateral and vertical relationships of deposits. Figure 11 shows facies migration during deposition of a clastic wedge.

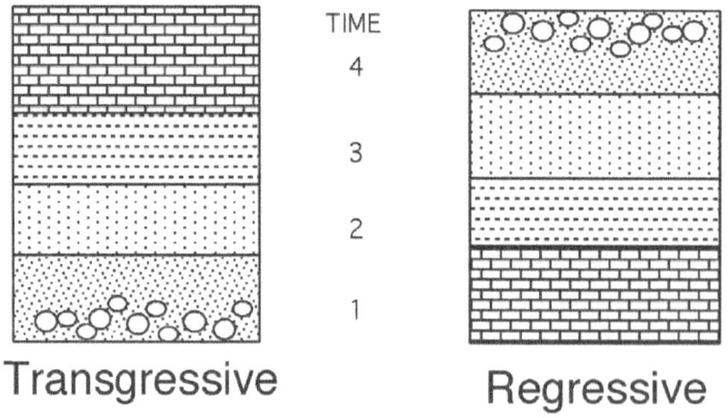

Figure 9 Transgressive sequence, with older conglomerate overlain by younger sandstone, then shale and limestone, indicating a rise in the sea level through time. Regressive sequence, with older carbonate strata overlain by younger shale, then sandstone and conglomerate, indicating a lowering of sea level through time.

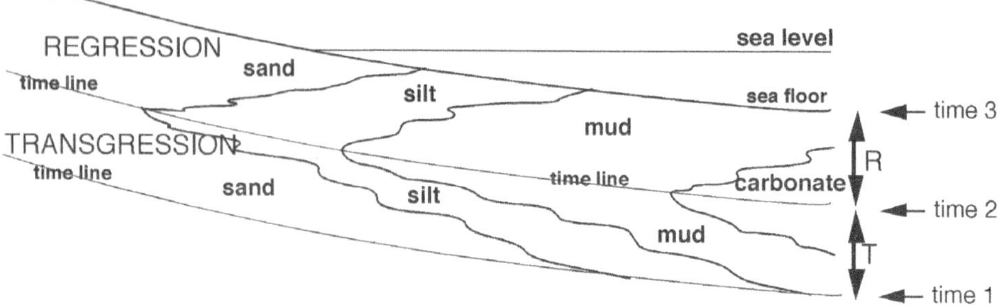

Figure 10 Lateral and vertical relations of depositional environments through time within a transgressive–regressive (T–R) cycle of deposition. The migration of depositional environments produces time-transgressive boundaries on lithologic units, shown by crossing of time lines.

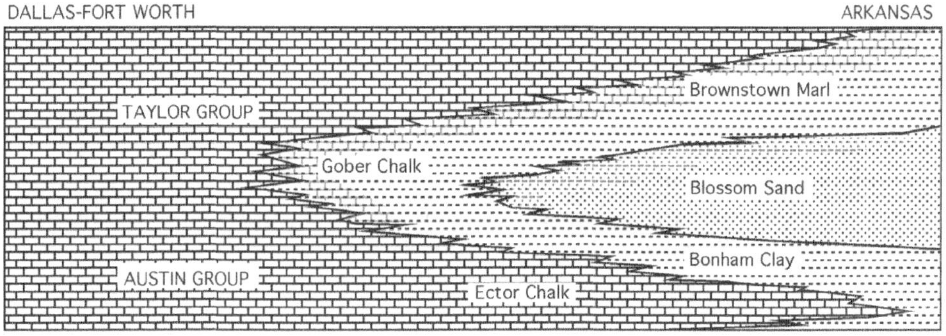

Figure 11 Clastic wedge of sediments deposited during the late Cretaceous in northeast Texas. This is a regressive–transgressive (R–T) cycle of deposition. Distance is plotted on the horizontal axis and time is plotted on the vertical axis, so time horizons correspond to horizontal lines. The true thickness of units may be different, because rates of sediment deposition vary along the transect. Near the sediment source area, there may be little space for sediment accumulation, so sediment is carried further into the basin, leaving only a thin layer near the source.

The vertical and lateral placements of lithologic facies in a T–R cycle of deposition is described in Walther's Law. It states:

> *For sediments deposited continuously during a transgression or regression, laterally adjacent facies will occur above or below the other in a conformable vertical sequence of deposits.*

This means that the vertical trend seen in the succession of deposits is similar to the lateral facies changes present in the area of deposition. If either the vertical or lateral arrangement of deposits is known within a cycle, a prediction of the succession of deposits in the other direction is possible. However, not all cycles are symmetrical. If the vertical trend is not similar to the lateral facies trend in an area, there may be an unconformity present that indicates a time of nondeposition or erosion or there was a sudden change in patterns of sediment transport and deposition.

Exercises

1. (RE-)LEARN THE GEOLOGIC TIME SCALE.

   It is necessary to learn all **eras** and **system/period** units of the time scale (Figure 12) and to learn all **series/epoch units of the Cenozoic Era**, as well as the two major units (**Archean** and **Proterozoic**) of the Precambrian. The important numerical dates to learn are the ages of boundaries between the eras and Precambrian units: 65 Ma, 251 Ma, 543 Ma, and 2,500 Ma.

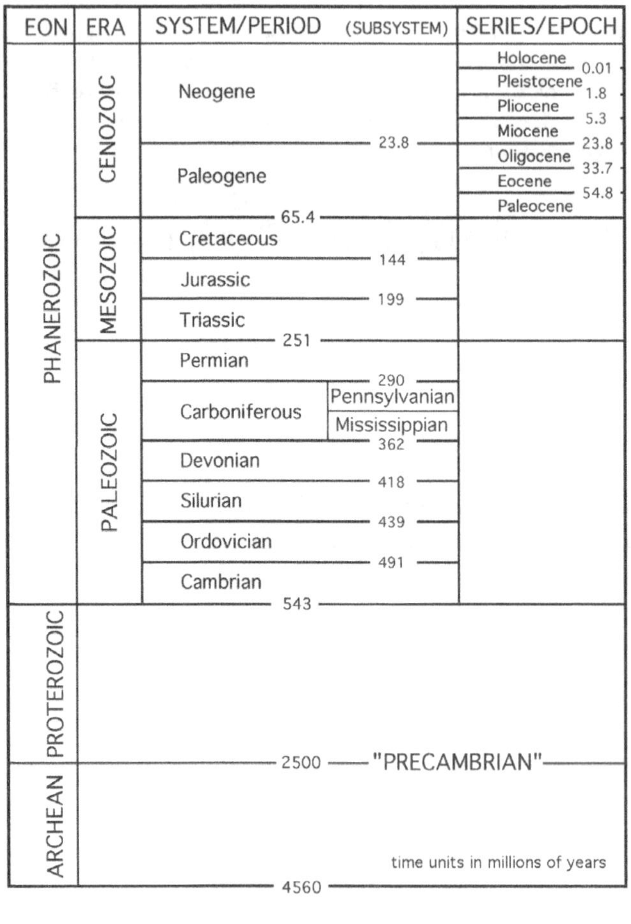

Figure 12 Geologic time scale.

Determining numerical ages

Numerical ages are determined by measuring the products of a time-dependent process and calibrating the results to a known standard. Radiometric dating is the most common approach to determining numerical ages.

Determining relative ages

To determine relative ages, identify the unconformities in each block and the succession of rock units of intervals separated by unconformities. Then look to see if the rock units (and their unconformities) have been folded or faulted or intruded. This will give a complete record to interpret.

2. Determine the relative ages of rock units in the diagrams below (Figure 13). To answer, write the letters of the rock units in a column to the right of the diagram, with the oldest units and events shown at the bottom and the youngest units and events at the top. At the proper places in the column, write in the geologic events (folding, intrusion, unconformity, etc.).

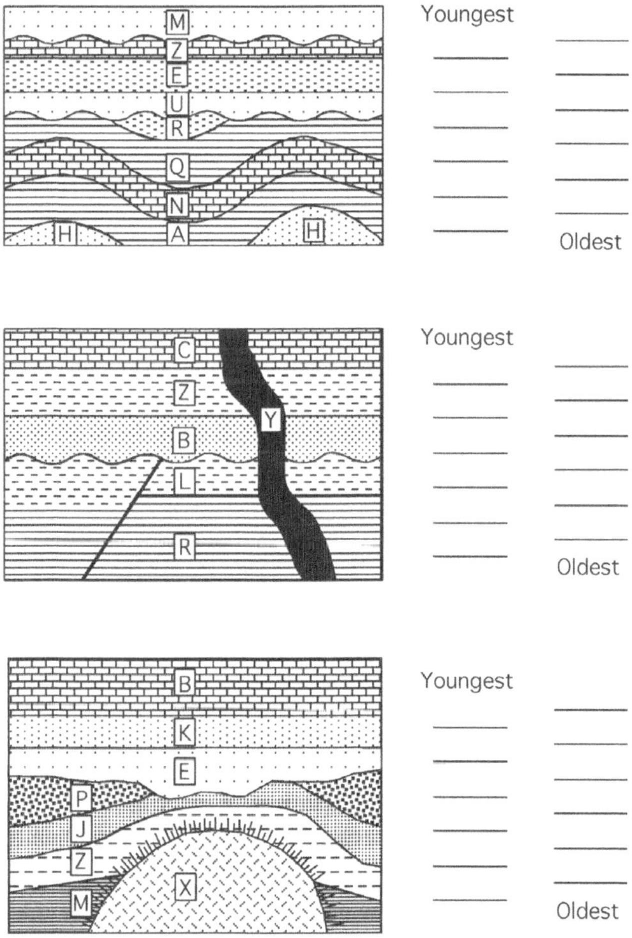

Figure 13 Relative ages.

Lithologic correlation of strata and transgressive–regressive cycles.

3. Draw lines of correlation between the units shown in the stratigraphic columns below (Figure 14), using lines drawn from levels of lithologic change (unit boundaries, which are called geologic contacts) in one column to other columns, to show the extent of units that were originally continuous across the area.

   A. How thick is the composite stratigraphic section?

   B. How many stratigraphic units are present in the composite section?

   C. A unit of conglomerate is present in sections 1 and 2, and is probably present in the other section as well. This is a water-bearing unit (aquifer) for the region and of great economic importance for homes and agriculture. Starting from the top of section 3, how deep would you have to drill to reach the conglomerate aquifer when drilling for water?

4. A. Correlate the lithologic units in the sections shown below (Figures 15, 16), drawing boundaries of the units. Boundaries of these units are lithostratigraphic surfaces.

   B. Identify the portions of the cycle of deposition that are transgressive deposits and the portions that are regressive deposits by labeling them with the letters **T** and **R**.

   C. Draw a dashed line along the horizon deposited during the time of maximum transgression, which represents the time of maximum flooding, labeling it **"maximum flooding horizon."** This is a chronostratigraphic (time) surface.

   D. Judging from the lithologies present, which type of sediment was deposited nearest to the land and which type of sediment was deposited farthest from the land?

   E. In which direction (north or south) was the nearest land?

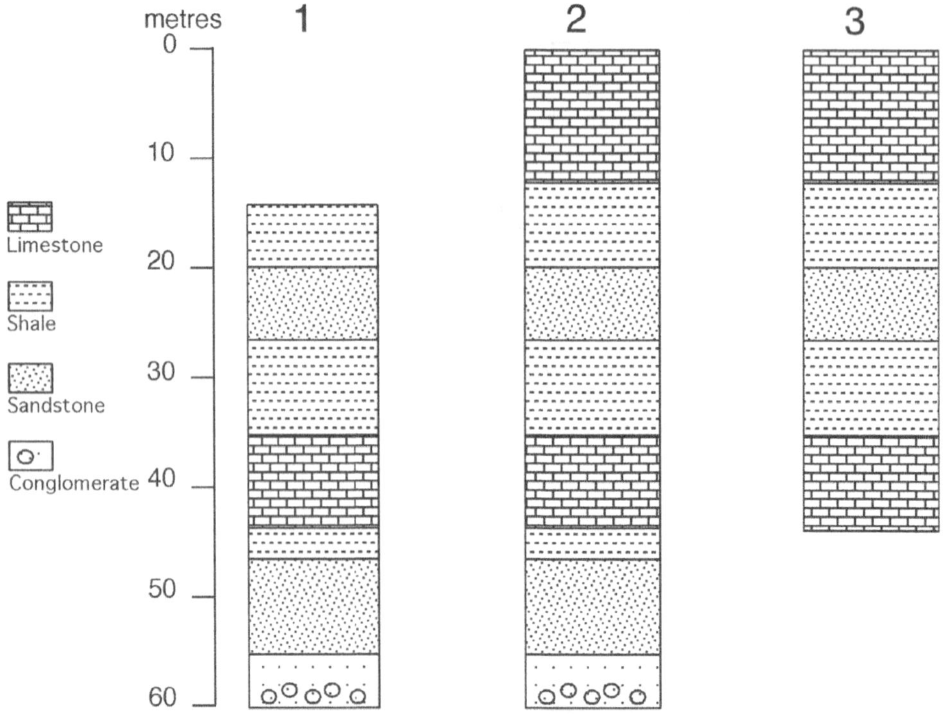

Figure 14 Correlation of stratigraphic sections.

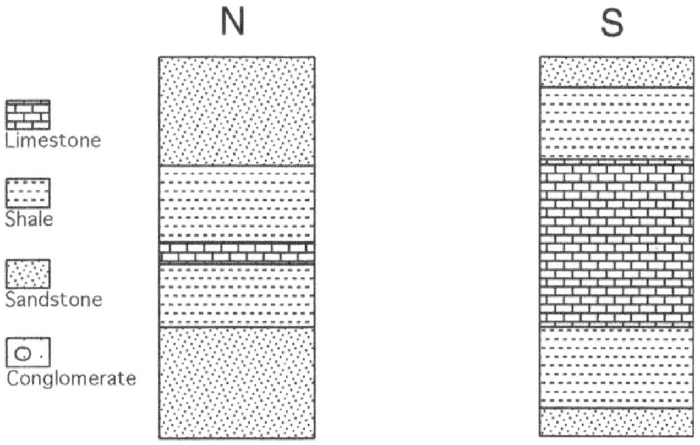

Figure 15 Correlation of stratigraphic sections.

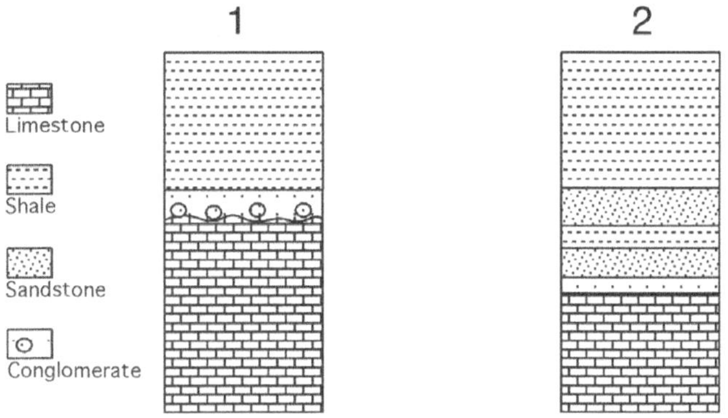

Figure 16 Correlation of stratigraphic sections.

5. Correlate the stratigraphic units in the two sections shown below and label the position of the unconformity.

   A. List the stratigraphic units (name the rock types) that are missing from the section with the unconformity.

   B. What type of unconformity is present in section 1?

6. Shown below is a composite cross section of geologic features of the Grand Canyon, set on a base in the northeast sector (north rim) of the canyon. The circled letters indicate geologic rock units and geologic structures such as faults and unconformities.

   A. List the geologic features (on lines to right of diagram) in the order they occurred, from oldest to youngest.

   B. The unconformity labeled TT has been offset by faulting from one of the unconformities shown in the center of the diagram. Does it correlate with unconformity QQ or unconformity SS?

   C. Is the unconformity labeled RR the same as unconformity QQ or unconformity SS?

   D. Identify the ages of rock units to the right of fault VVV.

   E. The fault labeled VVV comes to the surface in a canyon, where younger units are eroded away. However, there has been movement on the fault and that provides data on the age of faulting. Determine the age of faulting and direction of movement. State your answer in terms of younger than "X" and older than "Y."

   F. What feature of the contacts between the schist and gneiss complex (Unit N) and the granite complex (Unit O) allows you to tell the relative ages (older vs. younger) between them?

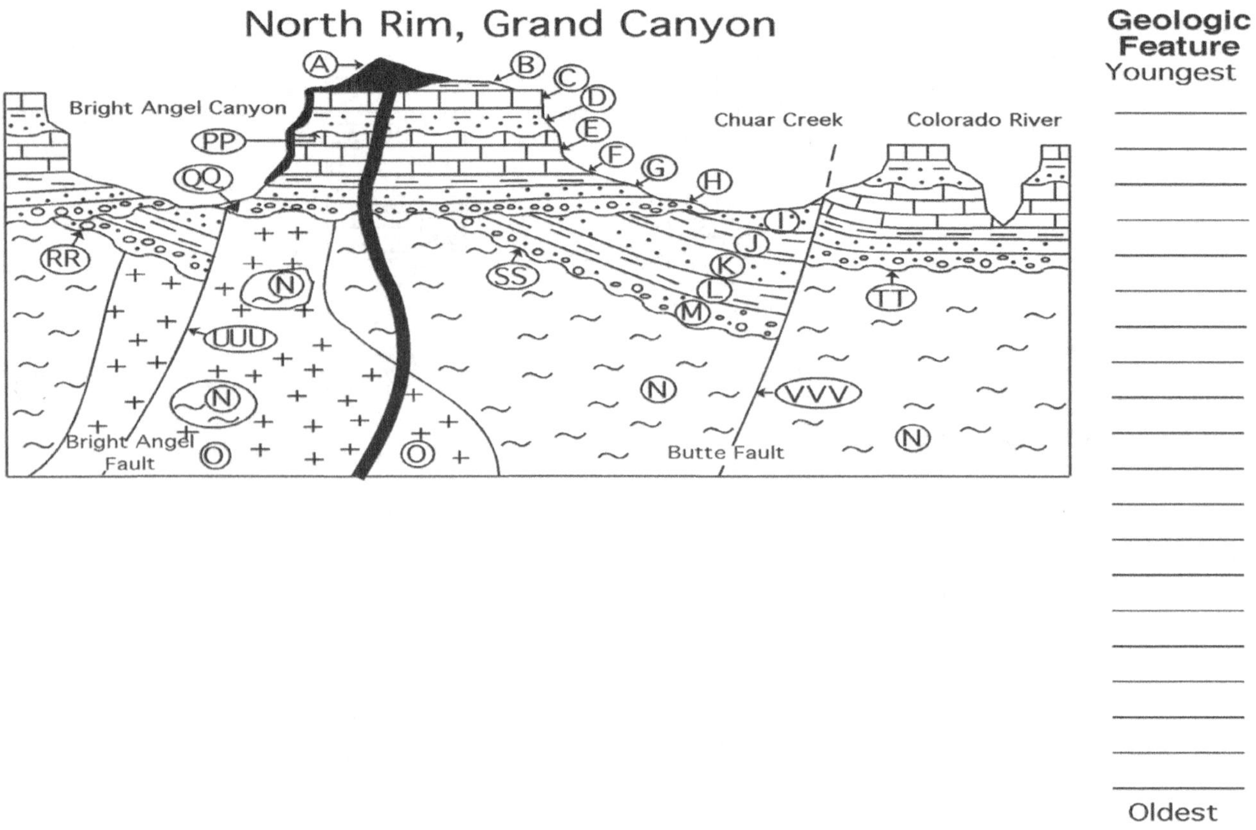

Figure 17 Schematic cross section of rock units in the Grand Canyon.

7. Write a geologic history of the Grand Canyon, based on the chart used in ordering geologic events in the area and using the geologic map of the Grand Canyon to guide your interpretation.

Procedure:

The purpose of writing a geologic history is to describe the depositional, erosional, and tectonic events that occurred in an area, written in a framework of progressive geologic time. Start with a description of the oldest rocks in the area and note their relationships to adjacent rock units. Are original contacts with other rocks preserved (depositional or intrusive), or is there a fault boundary, or an erosional, unconformable boundary? The nature of the contact will usually indicate the relative ages of rock units.

For example, the oldest rocks in the Grand Canyon are ancient crystalline rocks, intruded by magmas and overlain by sedimentary rocks. The first step is to determine if the metamorphic rocks or the intrusive rocks are the older of the two types. The next step is to determine relations between the crystalline rocks and sedimentary rocks. In the process of answering the question "Is the contact between the crystalline rocks and sedimentary rocks gradational, faulted, or erosional?" we see that there is a layer of coarse sand and gravel above the crystalline rocks, suggesting a nonconformity; therefore, we can determine a sequence of events consisting of metamorphism, intrusion, uplift, erosion, and transgression by ocean waters, requiring a long interval of geologic time. This process is repeated for interpreting the geologic history of younger units.

Look for lithologic trends in the deposition of sedimentary strata. If there are fining upward trends or vertical successions of depositional environments, such as sand to mud to carbonate, this suggests transgression. Look for systematic changes in depositional environments, such as changes from nonmarine to shallow water to deep water, and vice versa. Sedimentary deposits like oolitic limestone suggest shallow-water deposition and deposits of coal and red beds indicate nonmarine deposition. Depositional trends frequently are the products of T–R cycles. Describe the timing of any T–R cycles present and the types of sediments deposited during them.

At places where strata of certain ages are missing, a major unconformity must be present. Indicate the times (duration) of the exposure and erosion that produced the unconformity. If volcanic rocks are present, they may be intrusions into older rock or occur as flows that are interbedded with older rock below and younger rock above.

Cite the geologic evidence for each geologic event mentioned in the report.

Basin Analysis

Lab 7

Part 1

The Interior Seaway Depositional Basin

During the Cretaceous, North America had high mountains on the western part of the continent, lower mountains along the eastern edge, and a central lowlands area. Global sea level was 250 m higher during the mid-Cretaceous, covering the interior from the Gulf of Mexico to the Arctic Ocean in a large Interior Seaway. Mountain building and uplift were active in the west, creating highlands that shed large amounts of sediments into the western edge of the interior seaway.

The locations of 15 (1–15) stratigraphic sections are plotted on a base map. Thickness and lithology of strata in these sections are drawn on data sheets arranged in three transects: 1–5, 6–10, and 11–15. The number above the section corresponds with the location on the map. Lithology is shown with standard symbols shown in a legend. Using available data: correlate the lithologic units (formations), make a time correlation on transects (using radiometric dates, etc.), and determine the distribution of depositional environments in the Interior Seaway basin.

This exercise work assumes that the area is a single basin, faulting and folding have not disturbed the lateral relations of the sections, and sediment deposition was continuous and postdepositional erosion was insignificant. Lateral facies relationships are of the type shown below (Figure 1):

Procedures

1. **Correlate equivalent ages of strata** by drawing connecting age lines across the three transects using ash beds dates. **Mark lightly with pencil, so the underlying data are not obscured by lines.** Numbers on radiometric dates that are within 2 million years of each other are regarded as the same. **You can draw in five time lines across the sections.**

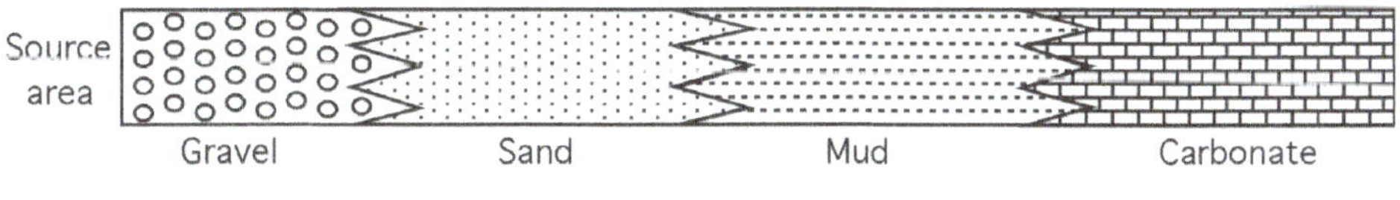

Figure 1 Facies model for sediments deposited in clastic wedges of the Interior Seaway.

2. Using a pencil, **draw boundaries (Figure 2) enclosing major lithologic units (lithofacies)** on the data sheets (Figures 3–5) and extend the boundaries from one section to another. Use a yellow-colored pencil to fill in units of sand and gravel conglomerate (use a light touch with the color fill), a blue pencil to color in the units of limestone, and leave the shale units uncolored. Formation boundaries usually cut extend across time lines (they are time transgressive). Use the concept of clastic wedges and transgressive–regressive (T–R) cycles of deposition to correlate formations from section to section.

3. Label each lithologic unit as fluvial, shorezone, inner (for inner shelf), or distal (for outer shelf or basin center).

4. Examine the age and lithologic correlations on transects and determine the times and extent of marine transgressions and regressions. Along the left and right margins of the three transects, **mark the level of cycle-center and cycle boundary horizons**. Then label the intervals between marks as T or R for transgressive or regressive. Transgressions occur where marine waters flood the land (or sea level rises) and regressions occur where the shoreline extends into the ocean basin (or sea level drops).

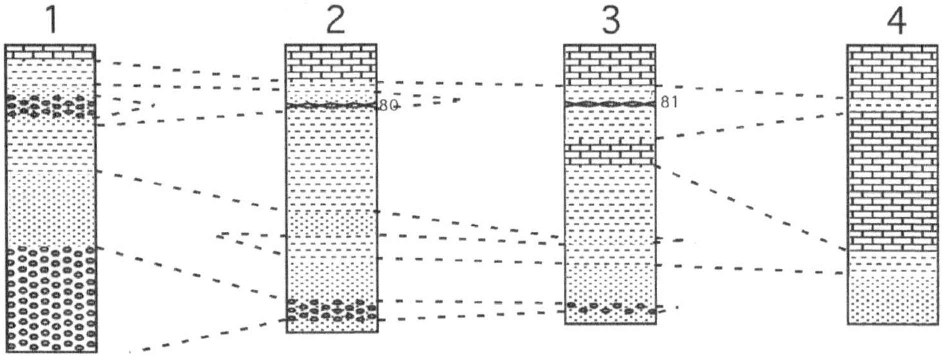

Figure 2 Example for lithologic correlation for work with stratigraphic sections.

5. For the 85 m.y. time line, construct a lithofacies map on the location map of the sections (Figure 6).

   A. Draw in the probable position of the shorelines. Since sands are typically deposited in beach zones, sandstone formations are indicators of shorelines in the facies model used for this exercise.

   B. Mark the location of mountain areas that were major source areas of clastic sediment to the basin, based on the distribution of conglomerate and sandstone.

   C. Use the distribution of basalts and volcanic ash beds to indicate where the greatest volcanic activity was located.

   D. Project the lithofacies boundaries south out to the Gulf of Mexico and north to the Canadian provinces.

   E. Draw a lithofacies map for the 70 m.y. time line and make the same set of interpretations.

6. Determine rates of sediment deposition (in m/million years) between 70 Ma to 75 Ma and 85 Ma to 90 Ma

	Section 1	Section 3	Section 5	Section 11	Section 15
Sediment rate					
70–75 m.y.					
85–90 m.y.					

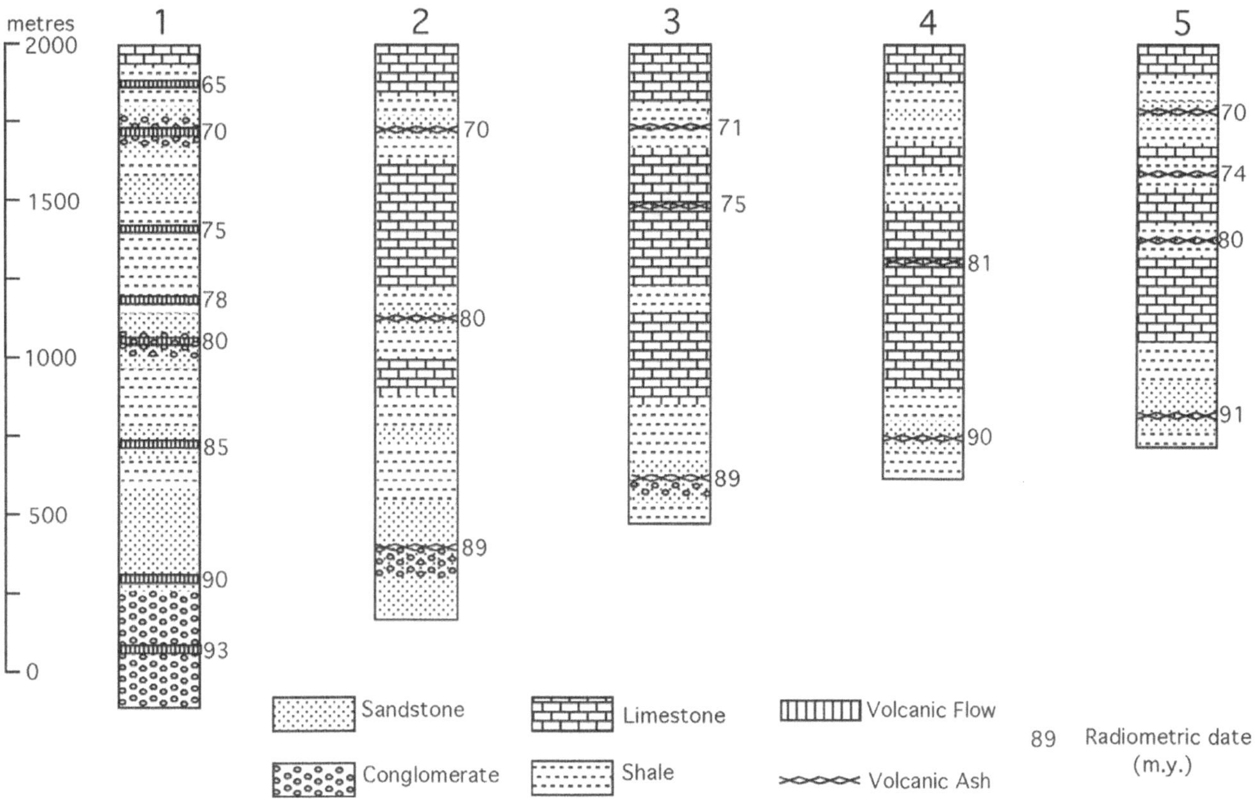

Figure 3 Sections 1–5 of Interior Seaway basin.

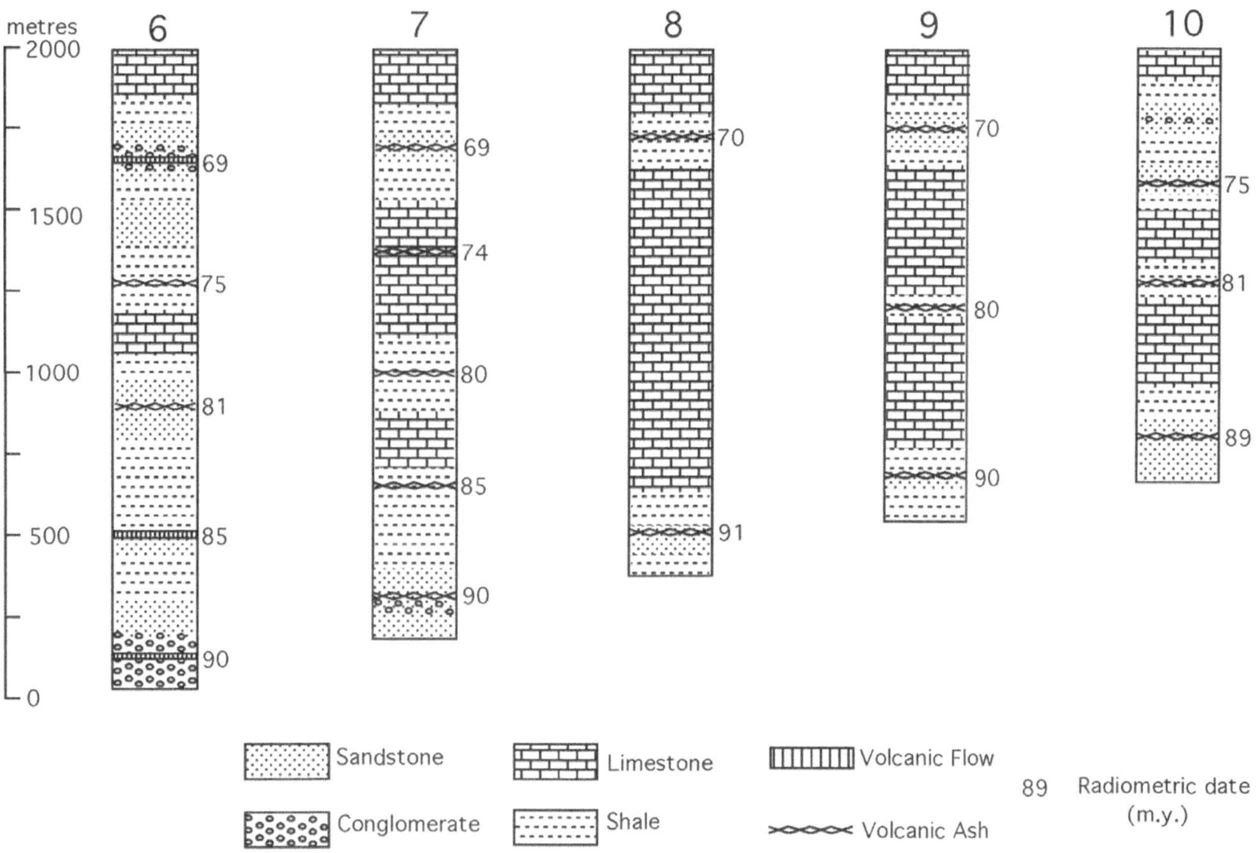

Figure 4 Sections 6–10 of Interior Seaway basin.

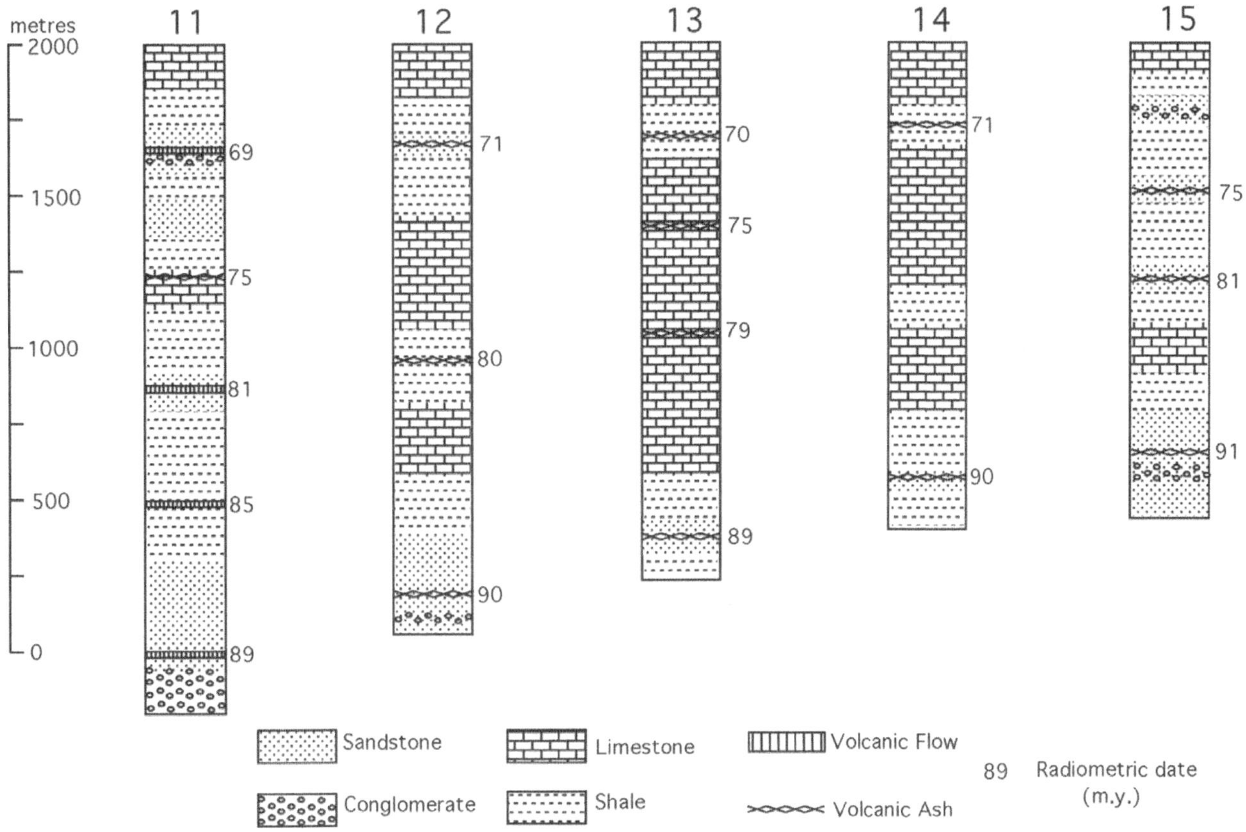

Figure 5 Sections 11–15 of Interior Seaway basin.

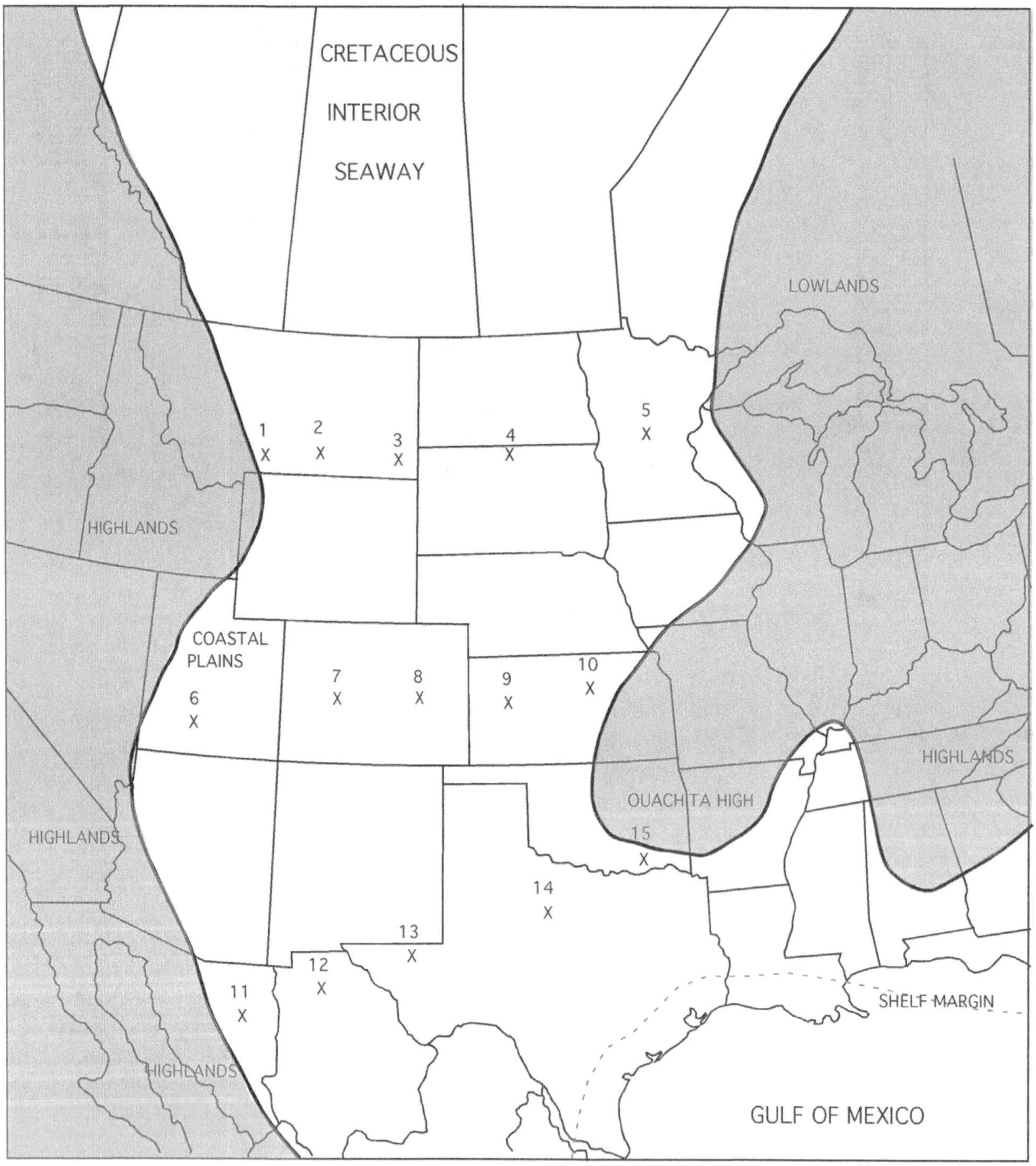

Figure 6 Location map of sections in Interior Seaway basin.

Part 2

Construction of Stratigraphic Cross Sections

In some areas near the tectonically active mountain ranges of the western zone, deformation has folded strata, eroded some of the section, and later deposition covered the strata. Using the information available from borehole records and from surface maps, construct a stratigraphic cross section and correlate lithologic units present in the subsurface.

Borehole Data. Two boreholes were drilled in the map area and the rock units encountered are summarized below. Plot the data in two columns on a sheet of paper, then draw correlation lines connecting units that occur in both sections, thus correlating rock units between the boreholes.

Borehole 1

Top

Unit 9	50 m of cherty limestone
Unit 8	50 m of gray shale
Unit 7	25 m of argillaceous limestone
Unit 6	25 m of coarse-grained sandstone
Unit 5	100 m of argillaceous limestone
Unit 4	100 m of oolitic limestone
Unit 3	100 m of gray shale
Unit 2	100 m of fine-grained sandstone

Bottom

Borehole 2

Top

Unit 4	25 m of oolitic limestone
Unit 3	100 m of gray shale
Unit 2	100 m of fine-grained sandstone
Unit 1	150 m of black shale

Bottom

Instructions for Completing Work

A. On the profile sheet (Figure 7), plot the vertical thickness and relative position of all rock units encountered in each borehole. Label each rock unit with its name (e.g., Unit 1).
B. Correlate the upper boundary and lower boundary surfaces of rock Unit 3 between the two boreholes by drawing lines connecting them from one borehole to the other.
C. Project the contact (common boundary) between Unit 1 and Unit 2 in Borehole 2 towards Borehole 1. The base of Unit 2 lies at the bottom of Borehole 2.
D. Placement of the contact between Unit 5 and Unit 6 can be determined by looking at the geology map. You can see that Unit 6 is a flat lying bed. Project the basal contact of Unit 6 laterally from Borehole 2.
E. To determine placement of other contacts on the cross section, use the geologic map to mark the places where the contacts of rock units intersect the profile of the ground surface. To do this, draw a line connecting point A with point A', then place the lower margin of the profile sheet against the upper margin of the map sheet and project a perpendicular line (a right angle) from the A–A' line to the ground profile line. Mark the point where the projection line crosses the ground profile.
F. Read the strike and dip for each bed from the geologic map, then use a protractor to mark the inclination of the beds at the ground surface, and then project the contacts of rock units down into the subsurface from the ground surface on the cross-section diagram. Connect with the segments of contacts drawn between boreholes 1 and 2 and extend all contacts to the left or bottom margin of the cross section.

Questions

8. Why is Unit 4 thinner in Borehole 2 than Borehole 1?

9. What is the name for the type of contact present between Units 5 and 6?

10. Outline a geologic history of events that have affected the strata in this area.

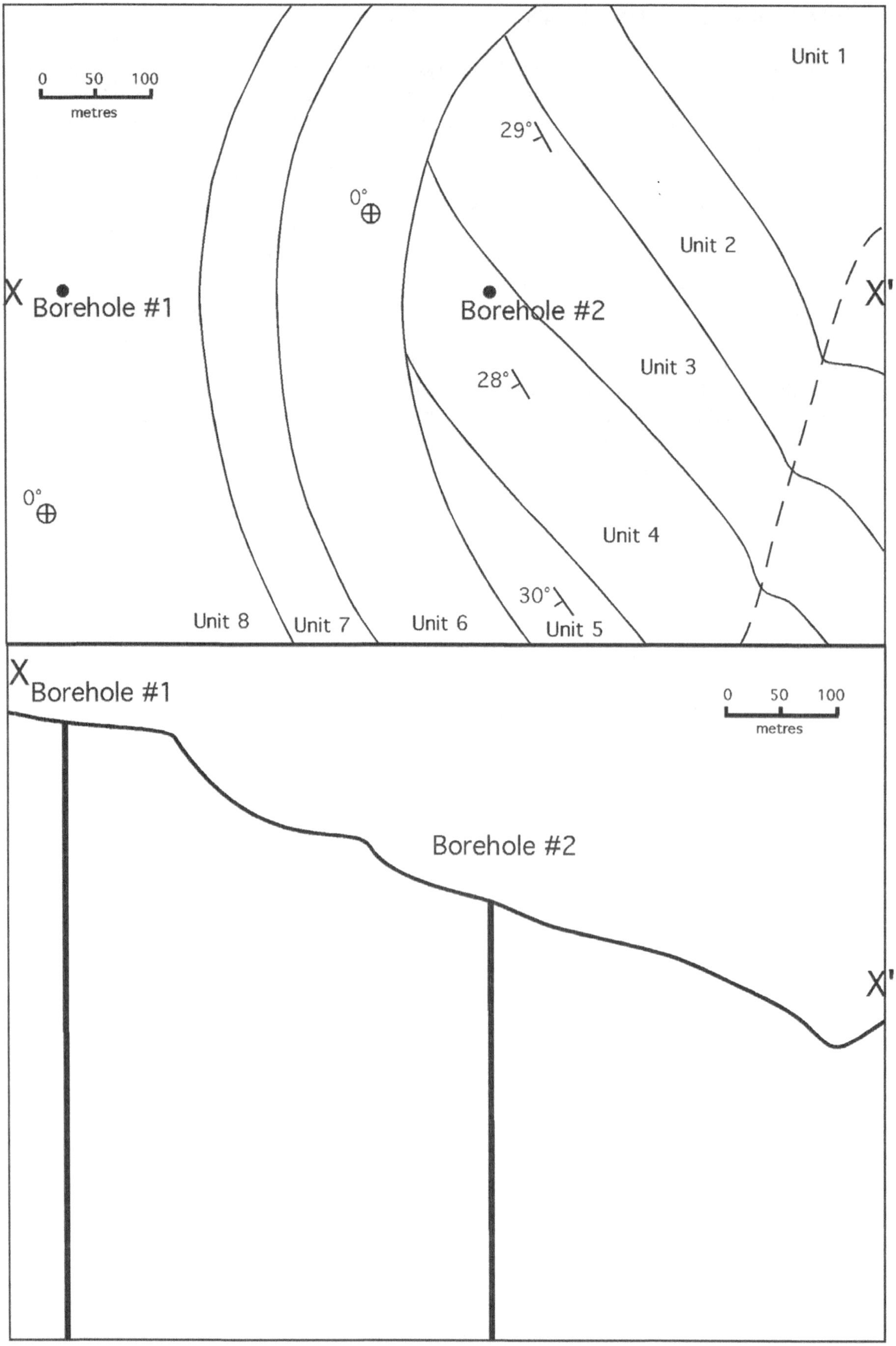

Figure 7 Geologic map view (top) and vertical cross section view (below) of the area.

Fossils and Biostratigraphy

Lab 8

Introduction

Fossils are the practical "clocks" of geology. Correlation by the use of fossils to determine contemporaneity of rock units is **biostratigraphic correlation**, or **biostratigraphy**. The principle of biotic succession states that each successive time interval of Earth history contains a unique assemblage of fossil species and therefore rock units can be dated by analysis of their fossil content, when compared to standard reference sequences. The fossil groups, which are the most important in marine biostratigraphy, are the **trilobites** (extinct arthropods), **conodonts** (extinct chordates), **graptolites** (extinct hemichordates), **ammonites** (extinct cephalopods), **fusulinids** (extinct larger protists), **coccoliths** (minute protist phytoplankton), **planktic forams** (protists), and **spores and pollen**.

Biostratigraphic Zonation

Biostratigraphy is the procedure in which fossils are used to determine relative age of sediments, following the **Principle of Biotic Succession.** Each species or group of species existed during a limited interval of geologic time, so its presence in a rock can be used to date that rock. The basic unit of biostratigraphy is the **biozone**, the interval of strata containing the species (or group of species). The fossils most useful for biozonation are planktic or pelagic organisms living in the open ocean that are widely distributed. Species that meet these criteria are commonly called index fossils.

Defining Biostratigraphic Zones

Defining biostratigraphic zones is a very simple procedure that begins with plotting ranges of species against a scale based on the stratigraphic column from which the fossils were collected. When this information is available, boundaries of ranges are defined on the end points of species ranges (Figure 1).

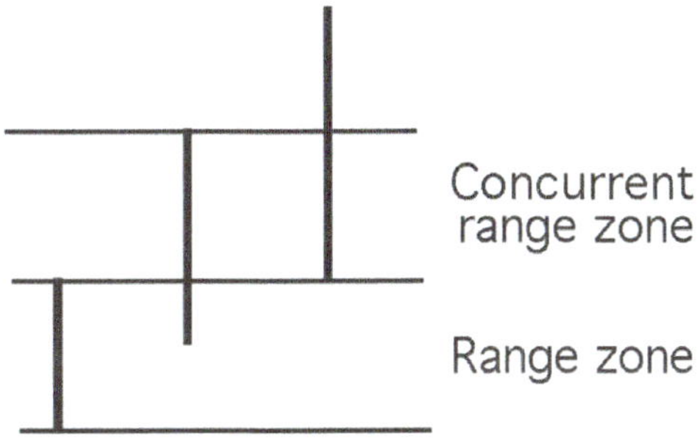

Figure 1 Example of biostratigraphic range zones.

Range zone: the body of strata containing the total range of the taxon in a stratigraphic sequence. This is the most basic type of biozone.

Concurrent range zone: the body of strata containing the overlapping ranges of two taxa.

Some Fossil Groups Used for Age Determination

Trilobites (Phylum Arthropoda)

Summary of Characters for Trilobites **Age Range: Cambrian–Permian:**

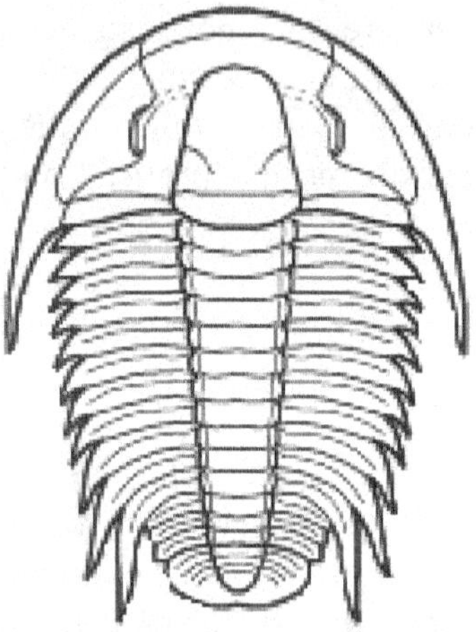	the group becomes extinct in the end-Permian mass extinction event From *Pictorial Guide to the Orders of Trilobites* by Sam Gon, http://www.trilobites.info/guidestart.htm. Copyright © 2015 by Dr. Sam Gon III. Used by permission.
Diagnostic Features:	segmented body, three-part division of skeleton, skeleton covering outside of body
Morphology:	trilobed, segmented exoskeleton, divided into head (cephalon), middle (thorax), and tail (pygidium), often with spiny projections on segments
Symmetry:	bilateral
Life Habits:	marine only; benthic crawlers and burrowers, detritus feeders and carnivores (the tiny agnostid trilobites were nekto-planktic inhabitants of the open ocean)

Learn to recognize trilobites. This extinct group was among the first groups of animals to acquire a mineralized skeleton: a calcite-impregnated external layer (an exoskeleton). The mineralized skeleton covered the upper parts of the body of the trilobite, including the side areas as well as some of the underside of the head region, leaving most of the underside of the animal unencumbered. The skeleton consists of many pieces (segments). Trilobites are arthropods that regularly shed their exoskeleton (molting) to allow for growth. Thus, each trilobite produced many empty, molted exoskeletons during its life. Trilobites are longitudinally divided into a trilobed condition with a raised middle band bordered by lateral zones and also a three-part division into head—cephalon, middle—thorax, and tail—pygidium. The body of the animal has many limbs, including numerous walking legs and pairs of antennae.

Ammonoids and Nautiloids (Class Cephalopoda; Phylum Mollusca)

Summary of Characters for Ammonoids **Age Range: Devonian–Cretaceous:**

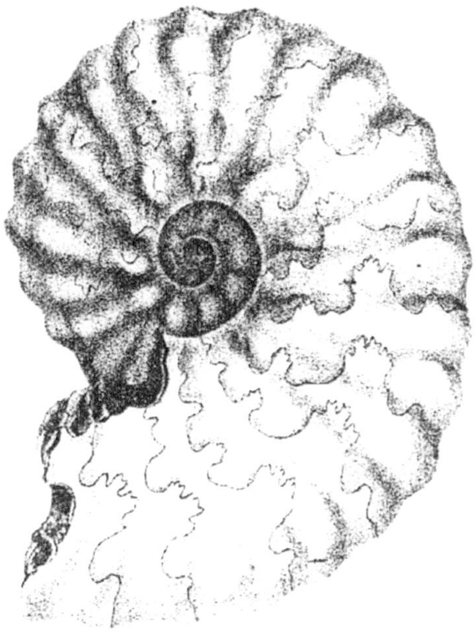	the group becomes extinct in the end-Cretaceous mass extinction event (Contrib. to Paleontology, C. White, 1880, U. S. Geol. Survev., pl. 18)
Diagnostic Features:	chambered shell, slender tube (siphuncle) passing through chambers beside the outer margin of whorl
Morphology:	coiled shell (at least as juveniles) with internal partitions that create chambers within the shell
Symmetry:	bilateral and coiled
Life Habits:	marine only; nekto-benthic to nektic carnivores

Learn to recognize ammonoids and nautiloids. Ammonoids and nautiloids are distinguished by having a chambered shell with a slender tube (siphuncle) passing through the chambers. Ammonoids are planispirally coiled and distinguished from nautiloids by having chamber partitions that are complexly folded. These show as **suture lines** on the inner wall of the shell. Ammonoids and nautiloids were mostly predators on other organisms. Ammonoids became extinct at the same time as the dinosaurs.

Suture Patterns

Learn how to recognize suture patterns. There is a progressive increase in complexity of suture patterns on shelled cephalopods. Nautiloids have the simplest sutures, consisting of a straight line around the chamber wall or a modest wavy line. With the evolution of ammonoids from nautiloids, the suture folding became pronounced (Figure 1). The next stage was secondary folding on the crests of some of the folds (Figure 2) and the most advanced stage was secondary and tertiary folding of all portions of major folds of the suture (Figure 3).

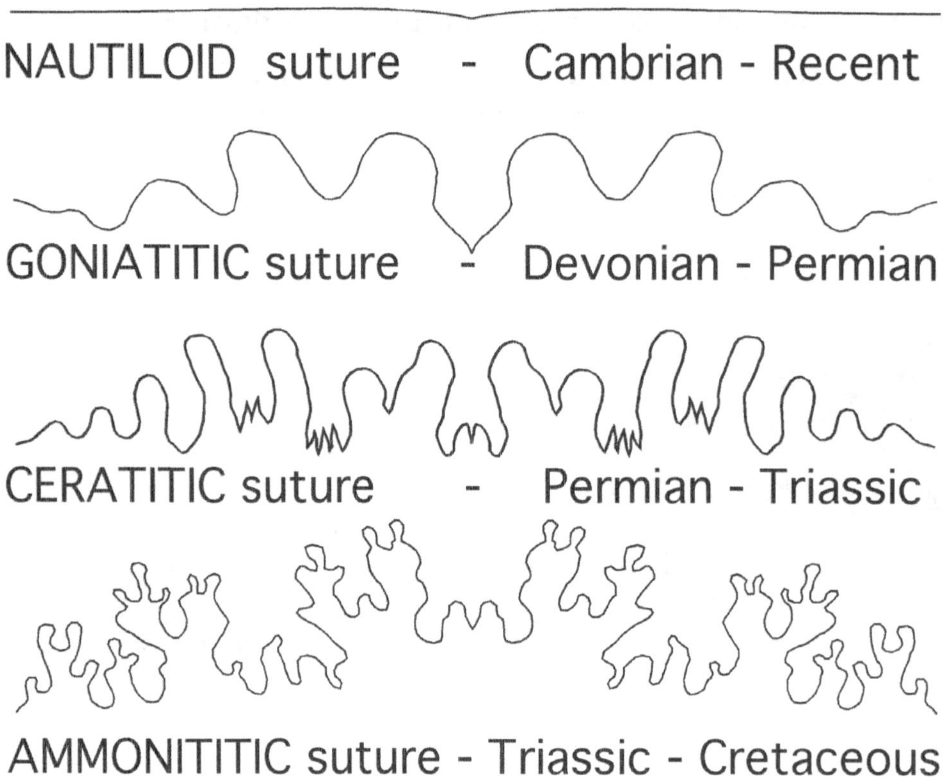

Figure 2 Nautiloid and ammonoid suture patterns.

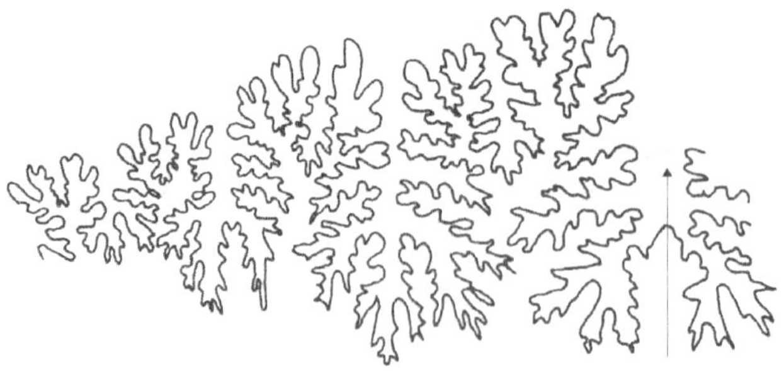

Figure 3 Example of very subdivided ammonoid suture pattern in Cretaceous ammonites.

Exercise Work

1. Examine the large cast reproductions (#254) of *Triarthrus* and determine the characters of these typical trilobites.

   A. How many legs are present under the cephalon (head)?

   B. How many legs are present under the thorax and pygidium (body portion behind head)?

   C. What form of locomotion is suggested by the form of the legs?

   D. Does it have pinchers to grasp prey animals and tear them to pieces for food?

   E. Could these animals fight against predators?

2. Trilobite #256–258 occurs as rolled up specimens, which is common for small trilobites and some had a nearly spherical morphology when enrolled.

   A. Suggest reasons for why these trilobites are rolled up and provide a functional interpretation of how a spherical form could provide protection from predators.

   B. These trilobites have large, elevated eyes with lenses that focus light, implying very good vision. Does having very good vision suggest a life habit of hiding and defensive behavior or an active, predatory life habit? Give reasons for your answer.

3. These distinctive fossils (*Isotelus*, #255; *Homotelus*, #253) are members of an Ordovician trilobite group with smooth, rounded body form and lacking projecting spines. They lived mostly in deeper water environments with mud substrate and the body form is probably an adaptation for burrowing through mud. Functional interpretation: How would a smooth body form make burrowing easier for these trilobites?

4. Compare the Early Cambrian trilobite *Olenellus* (#260) with the Late Cambrian trilobites *Elrathia* (#261) and *Asaphiscus* (#259). The form of *Olenellus* (#260) is typical of Early Cambrian trilobites, whereas the form of others is common among younger trilobites. Name two distinctive features of Early Cambrian trilobites such as *Olenellus*.

5. Examine the illustrations of trilobites shown in Part 2 (Biostratigraphic Zonation) that show average examples of early and middle Paleozoic age trilobites and examine specimens of *Homotelus* (#253), *Phacops* (#257), *Asaphiscus* (#259), *Olenellus* (#260), *Ditomopyge* (#264), and *Diacalymene* (#373).

   A. How does the average size of trilobites change during the Paleozoic: do they become larger or smaller?

   B. Projecting this trend to late Paleozoic, what can you predict about the average size of Permian trilobites?

   C. Trilobites lived from Cambrian to Permian, during the times of radiation of cephalopods, fish, and sharks (all predators). What changes occur in average trilobite morphology from Cambrian to Permian?

   D. Is there a reasonable correlation between the evolution of trilobites with that of predatory animals? If so, what is the correlation?

6. These tiny fossils (#376) are agnostid trilobites—a Cambrian–Ordovician planktic group of the open oceans. They are also known as "blind" trilobites, because the top skeleton lacks eye structures. Actually, the eyes of agnostids are located on the underside of the body, an area not covered with mineralized skeleton, so eye structures are not seen on these fossils. What characters of the agnostids are adaptations to a planktic life?

Ammonoids and Nautiloids

7. Using the diagrams of cephalopods suture patterns, determine the type of suture pattern and coiling pattern shown by the following specimens.

	Name of Fossil	Age	Suture Type	Coiling Style
A(#235)				
B(#241)				
C(#242)				
D(#244)				

8. Compare these specimens of uncoiled ammonoids (*Baculites*, #238) with straight nautiloids (*Isorthoceras*, #248; *Pseudorthoceras*, #250; and *Treptoceras* #251). How can uncoiled ammonoids be distinguished from straight nautiloids? List two criteria.

9. Specimens *Eutrephoceras* (#247) and *Metacoceras* (#245) are coiled nautiloids. How can they be distinguished from coiled ammonites?

10. Examine ammonoid specimen #237, which has a siphuncle that is darker than the rest of the shell. Is this located in the same part of the chamber as the siphuncle of the modern *Nautilus* shell (#249)?

11. Specimens *Mortoniceras* (#237), Goniatite (#240), and *Ceratites* (#242) show average sizes for Paleozoic and Mesozoic ammonoids.

    A. How does the average size of ammonoids change from the Devonian to the Cretaceous: do they become larger or smaller near the end of their geologic range?

    B. Is this evolutionary size trend the same as the size trend you determined for trilobites?

    C. Very large ammonites, up to 3 m in diameter, are known from some formations. What age would you assume for these very large ammonites?

Part 2

Biostratigraphic Zonation

12. On the chart below are geologic age ranges of several common trilobites as observed in a stratigraphic section.

    A. Draw in the boundaries of all the trilobite zones you can define. Label each zone with the names of the trilobites and identify it as a **range zone** or a **concurrent range zone**.

    B. Along the right side of the chart, mark the boundaries of geologic periods of the Paleozoic to show the geologic time associated with each trilobite zone.

The trilobites have geologic age ranges of:

Phillipsia:	Carboniferous–Permian
Phacops:	Silurian—Devonian
Cryptolithus:	Ordovician
Agnostus:	late Cambrian–early Ordovician
Elrathia:	late Cambrian
Olenellus:	early Cambrian

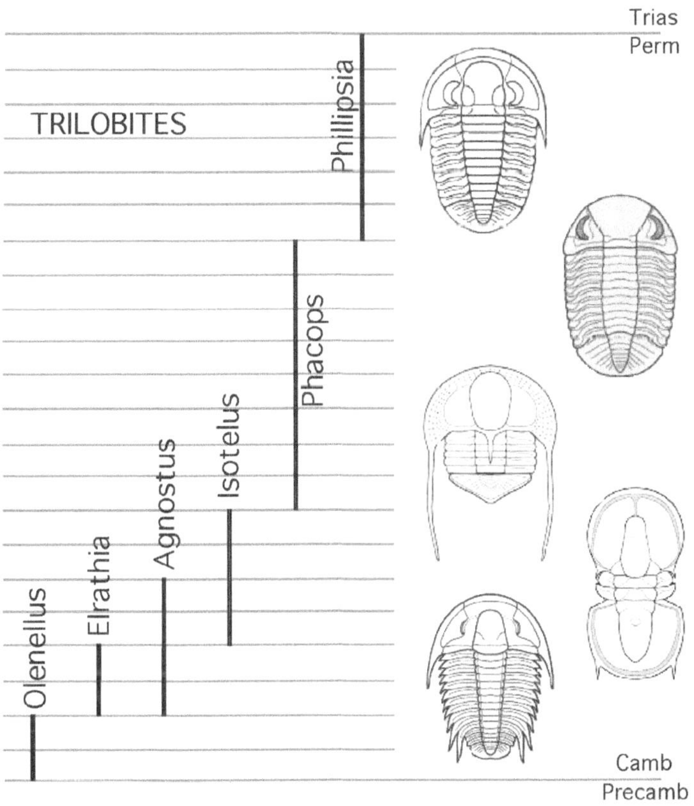

Trilobite drawings from *Pictorial Guide to the Orders of Trilobites* by Sam Gon, http://www.trilobites.info/guidestart.htm. Copyright © 2015 by Dr. Sam Gon III. Used by permission.

Figure 4 Age ranges of important trilobites.

Part 3

Using Fossils for Age Determination—Rattlesnake Creek, Gregwebb County

During the 1800s settlement of Texas, geological survey teams investigated Gregwebb County of central Texas for mineral and other deposits. A survey crew described the formations exposed along Rattlesnake Creek and collected fossils from the formations. No minerals were found, but the area was evaluated for clay to make brick, coal for fuel, and limestone or sandstone suitable for building stone. Evaluate the results of the survey. To start, determine the age of strata along Rattlesnake Creek, using samples collected by the survey team.

Exercise Work

13. Examine fossil samples and determine the geologic age of strata containing them. Samples consist of individual fossils or fossiliferous rock. Using the booklet "Common Fossils of Gregwebb County," identify the fossils and determine the age of the rock unit by the known age of the fossil or the overlap in age ranges of two or more fossils.

    Only six samples are available and one of these has no locality number. The samples are from sites 3, 7, 8, 9, 10, and "unknown." Samples 8 and 9 contain the same fossil and will be the same age.

Sample	Fossil name	Age range	Rock unit age
#3	 		Geologic time period:
#7			

Sample	Fossil name	Age range	Rock unit age
#8, #9	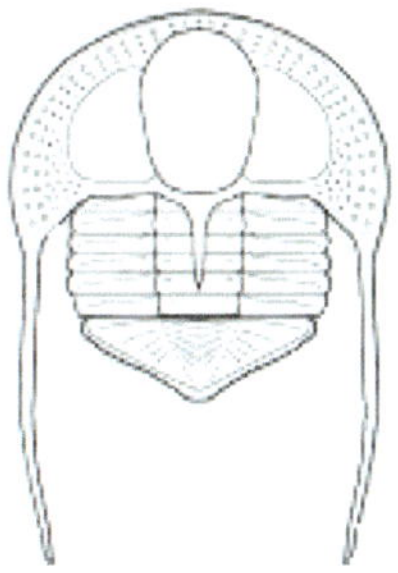 		
#10	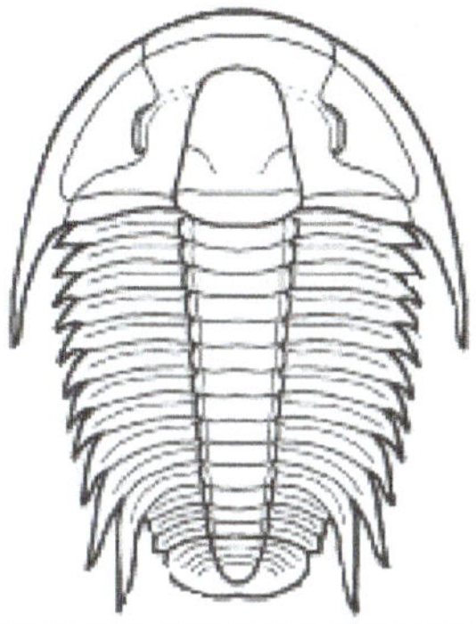		
#???	 		

Trilobite drawings from *Pictorial Guide to the Orders of Trilobites* by Sam Gon, http://www.trilobites.info/guidestart.htm. Copyright © 2015 by Dr. Sam Gon III. Used by permission. Photographs by T. Yancey.

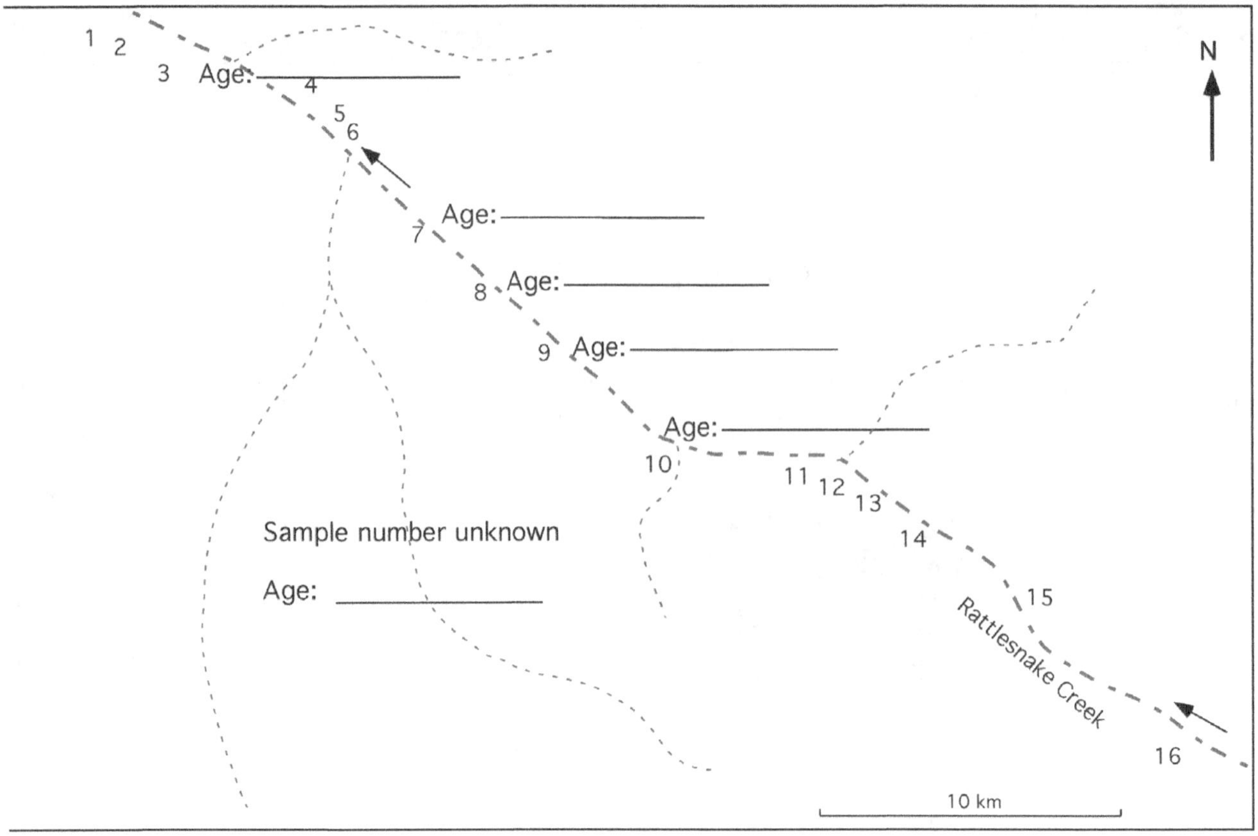

Figure 5 Sample collecting sites on Rattlesnake Creek.

14. On the map above, write in the geologic age of each sample on the line shown beside the collection site. With the ages of strata determined, a preliminary interpretation of the geology of the area can be made.

15. Do the geologic ages along Rattlesnake Creek indicate a simple progression from older to younger formations?

16. The arrows along Rattlesnake Creek indicate the direction of water flow, showing that the lowest elevation is to the northwest and the highest elevation is to the southeast. Do the oldest strata lie at the lowest elevation, consistent with an undisturbed, superposition set of strata?

Geological Exploration: The Geology of Gregwebb County, Texas: Part 1

Lab 9

Project Work

The geological survey of Gregwebb County, Texas, reported the occurrence of coal-bearing strata in the area. The deposit was not developed, but with the recent rise in oil and coal prices, interest in the resource has increased. A company has acquired the field notes, some field collections, core samples, and descriptions of cores that were drilled during an earlier investigation. You have been hired to look at these notes, descriptions, and collections to interpret the geology and evaluate the possibility of energy resources in the lease area.

Age Determination of Rocks and Strata in Gregwebb County, Texas

1. The first task is to determine the geologic ages of field samples by identifying the fossils in each sample. Determine the ages of samples 1–29 using the key taken from the book *Common Fossils of Gregwebb County*. Many samples contain the same combination of fossils and will have the same age. Write the age for each sample in the worksheets below. Two rocks contain igneous minerals that can be dated by radiometric dating; calculate dates for each sample.

2. Determine ages for fossils recovered from the three cores drilled in the area and list them in the worksheets below. Identify the age of all fossiliferous samples in the cores. Infer the ages for rock units lacking age-diagnostic fossils.

   Procedure:

   If the sample contains **one** fossil species, the age determination corresponds to the age range of that species: for example, a sample contains *Elrathia* (Camb)—therefore the age of sample is Cambrian.

   If the sample contains **more than one** fossil species, the age determination corresponds with the age of the species with the shortest age range: for example, a sample contains *Cyptolithus* (Ord) and *Halysites* (Ord-Sil)—therefore the age of sample is Ordovician (shown the presence of short-ranging *Cryptolithus*).

Sample	Fossil name	Fossils present	Rock unit age
#1	Volcanic ash present in rock unit.	1. Name the type of rock. 2. Calculate age of the rock. K-feldspar crystals contain 12.5% of original radiogenic isotope of a mineral with a half-life of 105 million years.	Numerical date: Geologic time period:
#2 and #29			Geologic time period:

Sample	Fossil name	Fossils present	Rock unit age
#3, #4, #13, and #28	Age determined in previous exercise. (#4, #13, and #28 same age as #3)		Age: Early Carboniferous
#5, #12, and #27	No fossils.	Name the type of rock.	Geologic time period: (Same age as Sample 3)
#6, #11, #21, and #26			Geologic time period:
#7 and #25	Age determined in previous exercise. (#25 same age as #7)	Fill in answer from previous exercise.	Age: Silurian
#8, #9, and #22	Age determined in previous exercise. (#22 same age as #8 and #9)	Fill in answer from previous exercise.	Age: Ordovician
#10	Age determined in previous exercise.	Fill in answer from previous exercise.	Age: Cambrian
#14	No fossils.	Name the type of rock.	Geologic time period: (Same age as Sample 15)
#15, #17, #19, and #20	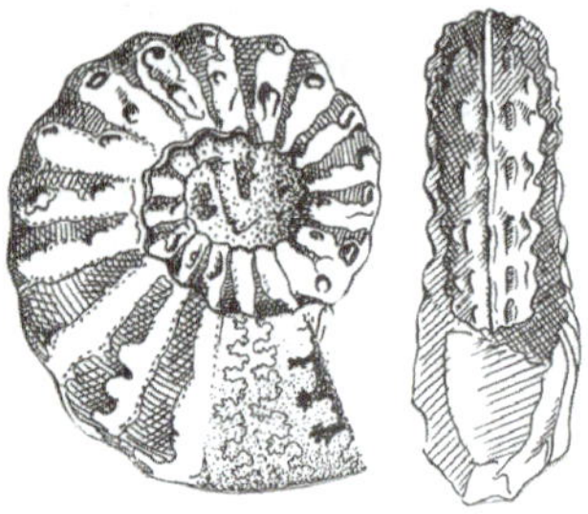		Geologic time period:
#16 and #18	Age determined in previous exercise. (#18 same age as #16)	Fill in answer from previous exercise.	Age: Eocene
#23 and #24			Geologic time period:

Age Determination of Fossils in Core Samples, Gregwebb County, Texas

Core 1

Sample	Fossil name	Fossils and Age range	Rock unit age
Unit 7			Geologic time period:
Unit 6	No fossils collected.	Name the type of rock.	Geologic time period: (Same age as Unit 7)
Unit 5			Geologic time period:
Unit 4	No macrofossils collected. Conodont microfossils indicate Silurian age.		Age: Silurian
Unit 3			Geologic time period:
Unit 2	No fossils. (sample lost)		Geologic time period: (Same age as Unit 3)
Unit 1	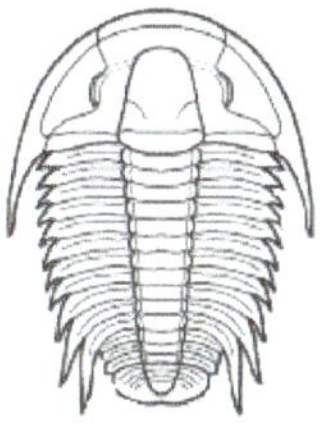		Geologic time period:

Trilobite drawings from *Pictorial Guide to the Orders of Trilobites* by Sam Gon, http://www.trilobites.info/guidestart.htm. Photographs by T. Yancey.

Core 2

Sample	Fossil name	Fossils and Age range	Rock unit age
Unit 5	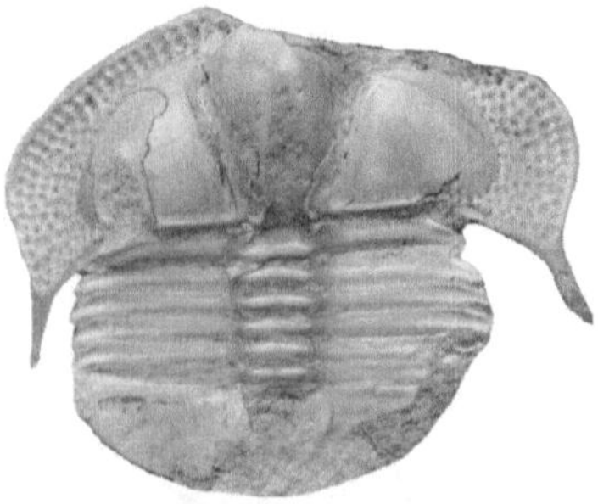		Geologic time period:
Unit 4	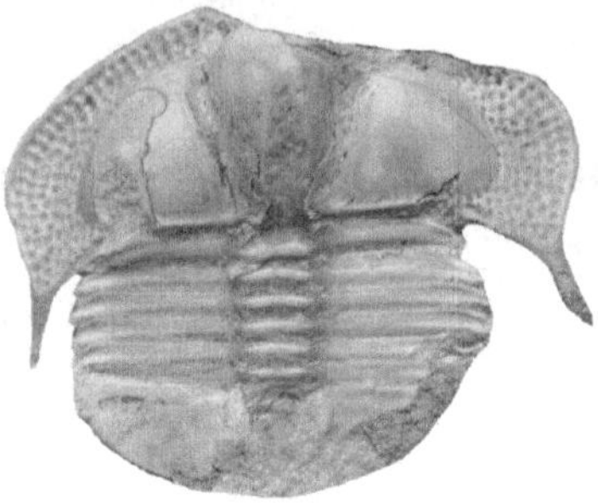		Geologic time period:
Unit 3	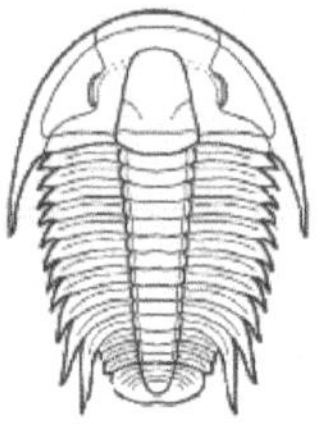		Geologic time period:
Unit 2	No fossils collected.	Name the type of rock.	Geologic time period: (Same age as Unit 3)
Unit 1	Igneous rock.	Calculate the age of the rock. Zircon crystals contain 37.5% original radiogenic isotope of a mineral with a half-life of 700 million years.	Numerical date: Geologic time period:

Trilobite drawings from *Pictorial Guide to the Orders of Trilobites* by Sam Gon, http://www.trilobites.info/guidestart.htm. Photographs by T. Yancey.

Core 3

Sample	Fossil name	Fossils and Age range	Rock unit age
Unit 11	 		Geologic time period:
Unit 10	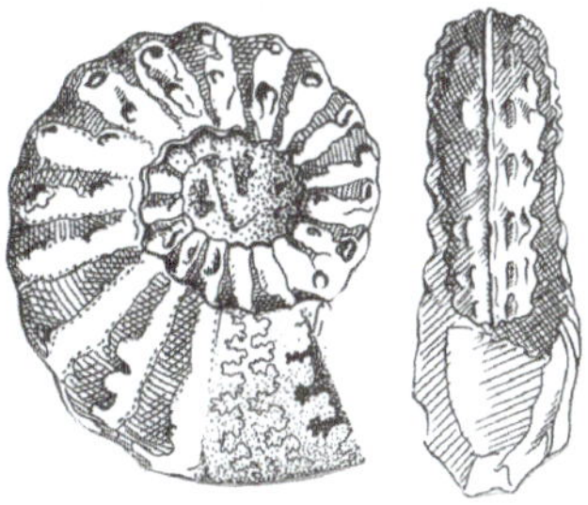		Geologic time period:
Unit 9	No fossils collected.	Name the type of rock.	Geologic time period: (Same age as Unit 10)
Unit 8	No fossils collected.		Geologic time period: (Same age as Unit 7)
Unit 7			Geologic time period:
Unit 6			Geologic time period:
Unit 5	No fossils collected.	Name the type of rock.	Geologic time period: Age determined by enclos-ing strata.

Sample	Fossil name	Fossils and Age range	Rock unit age
Unit 4			Geologic time period:
Unit 3	 		Geologic time period:
Unit 2	No fossils collected.	Name the type of rock.	Geologic time period: Same age as Unit 3.
Unit 1			Geologic time period:

Determination of the Stratigraphic Section in Boreholes, Gregwebb County, Texas

3. A major change in lithology occurs between Units 1 and 2 in core #2. What is the nature of the contact between them?

4. The next task is to determine the rock units present in the area. Plot the thickness of rock units listed in the core reports onto the work sheet below, using thickness listed in the reports. Plot units downward from the ground surface. Thicknesses of units reported in core descriptions have been corrected for variations in dip. Use standard lithologic symbols for each rock type (Figure 1) when plotting units on the work sheet (Figure 2).

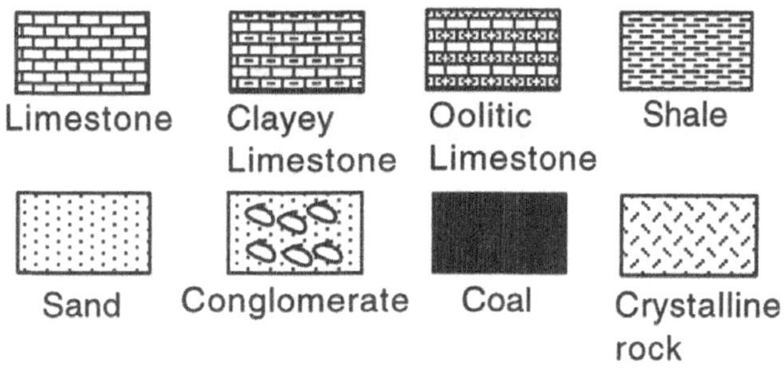

Figure 1 Lithology patterns used to show rock type.

5. Correlate the rock units present in the cores using age determination as the basis for correlation: mark system boundaries (e.g., between Cambrian and Ordovician) on sides of the sections, then draw correlation lines at rock unit boundaries from core to core.

6. The sections cannot be correlated by drawing straight horizontal lines; **so the strata must be deformed by faulting or folding or be eroded away in some places**. Make the unit contacts fit a realistic structure interpretation.

7. Using data from the core reports, construct a single, composite stratigraphic section for the area onto the composite section scale (Figure 3). **Read the section titled "Producing a Composite Stratigraphic Section."**

8. Using the composite stratigraphic section, write up a brief (one page length) geologic history of the area based on your core interpretations. Mention the following:

   A) the stratigraphic succession of depositional environments (as shown by changes in lithology),

   B) times of transgression and regression,

   C) unconformities, and

   D) times of uplift and erosion.

West Texas Energy Company Report 276—Gregwebb County Core Descriptions

CORE 1: This core was drilled in the northwestern area of the lease, beside the dry stream bed. The coring reached a depth of 830 m before drilling had to be suspended due to flash flooding.

Unit 7	150 m	of gray shale
Unit 6	50 m	of oolitic limestone
Unit 5	200 m	of argillaceous limestone, with irregular bedding
Unit 4	150 m	of gray shale
Unit 3	150 m	of interbedded limestone and shale, thin bedded
Unit 2	100 m	of limestone
Unit 1	30 m	of gray shale

CORE 2: This core was drilled southeast of the first borehole, near the creek. It reached a depth of only 490 m before granite was encountered. Drilling was terminated when 10 m of granite was penetrated.

Unit 5	150 m	of interbedded limestone and shale, thin bedded
Unit 4	100 m	of limestone
Unit 3	200 m	of gray shale
Unit 2	30 m	of coarse-grained sandstone
Unit 1	10 m	of granite

CORE 3: This core was drilled near the creek in the southeastern part of the lease. The section was cored to a depth of 1150 m, when the depth limit of the small rig was reached and drilling was suspended.

Unit 11	100 m	of interbedded sandstone and shale
Unit 10	90 m	of chalky, micritic limestone
Unit 9	10 m	of coarse sandstone, with conglomerate at base
Unit 8	50 m	of red sandstone and siltstone
Unit 7	50 m	of sandy limestone
Unit 6	180 m	of interbedded sandstone and shale, thin bedded
Unit 5	20 m	of coal with volcanic ash layer
Unit 4	100 m	of interbedded shale and limestone, thin bedded
Unit 3	300 m	of black shale
Unit 2	50 m	of oolitic limestone
Unit 1	200 m	of gray limestone, thick bedded

Use these patterns when plotting rock units.

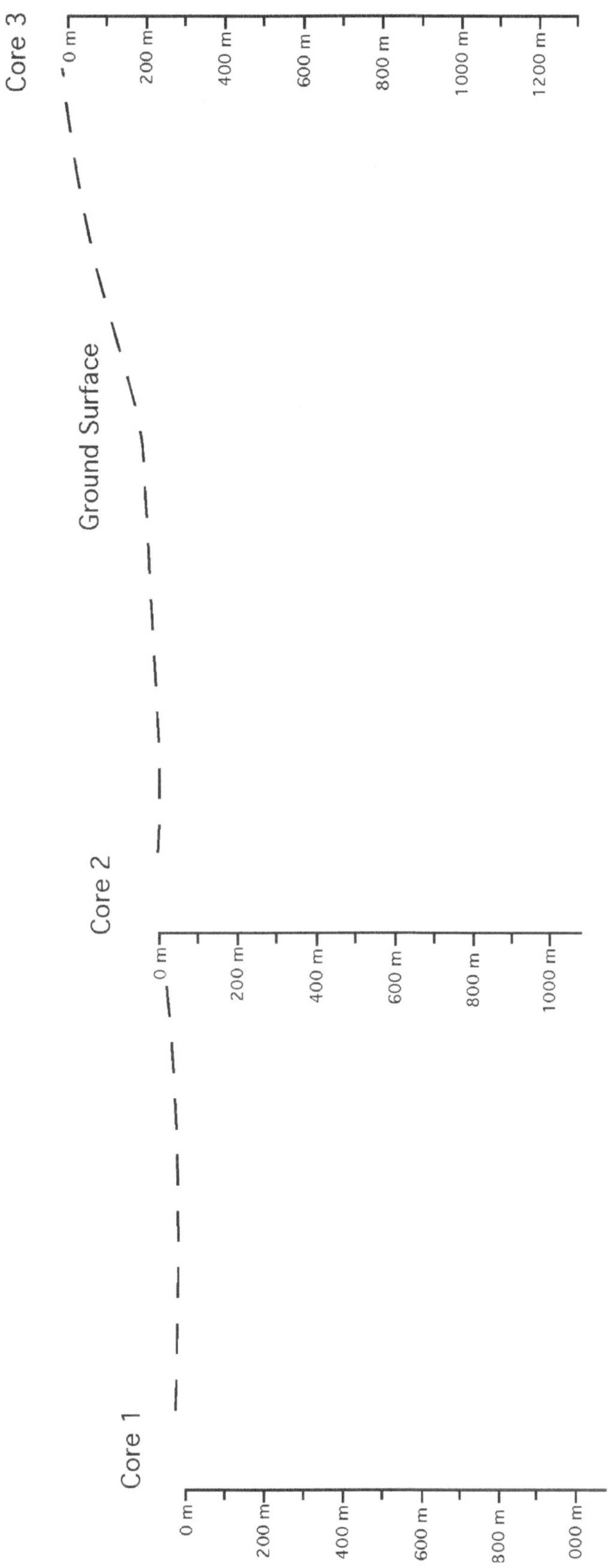

Figure 2 Work sheet for plotting rock units.

Producing a Composite Stratigraphic Section

A composite stratigraphic section is one that shows all the lithostratigraphic units present in an area with all units stacked in a pile. It is used as a standard of reference, so that missing intervals (lost by erosion, or faulting) or duplicated section (repeated by faulting) can be identified. The composite section is used to identify structure and assess geologic history of the area.

First, identify a unit that appears in two or more sections; plot that interval on a vertical scale in Figure 3, **showing relative thickness of units and lithologic character**. Use a vertical scale of 1 cm = 100 m and use lithology patterns to indicate the sediment type.

Second, identify intervals of overlying strata or underlying strata and plot them on the composite section scale. Repeat the process until all units in local sections are shown on the composite section.

Check to see that no units are missing and that none are duplicated on the composite. This is your reference section.

Writing a Geologic History

The purpose of writing a geologic history is to describe the depositional, erosional, and tectonic events that occurred in an area, written in a time framework. Start by describing the oldest rocks in the area and note their relationships to adjacent rock units. **Ask the question**: are original contacts with other rocks preserved (depositional or intrusive, etc.), or is there a fault boundary, or an erosional, unconformable boundary? The contact will indicate the relative age of rock units.

For example, the oldest rocks are crystalline rocks, overlain by sedimentary rocks. In the process of answering the question "Is the contact gradational, intrusive, faulted, or erosional?" note that there is a layer of coarse sand above granite, suggesting a nonconformity—therefore we can determine a sequence of events consisting of intrusion, uplift, erosion, and transgression by ocean waters, requiring a long interval of geologic time. The process is repeated for younger units.

Look for lithologic trends in the deposition of sedimentary rocks. If there are upward-fining trends or vertical successions of depositional environments such as sand-to-mud-to-carbonate, this suggests transgression. Look for systematic changes in depositional environments, such as changes from nonmarine to shallow water to deep water, and vice versa. Sedimentary deposits such as oolitic limestone suggest shallow-water deposition and deposits of coal and redbeds indicate nonmarine deposition. Describe the timing of the transgressive–regressive cycles and the types of sediments deposited within cycles.

At levels where a geologic period is missing, a major unconformity must be present. Indicate the times (duration) of the exposure and erosion that produced the unconformity. If volcanic rocks are present, they may be intrusions into older rock or occur as flows interbedded with older rock below and younger rock above. If the sedimentary rocks are **not** flat lying, there has been some structural change. Identify the type of structure and date the time of deformation.

Cite the geologic evidence for each geologic event mentioned in the report and the geologic time of occurrence.

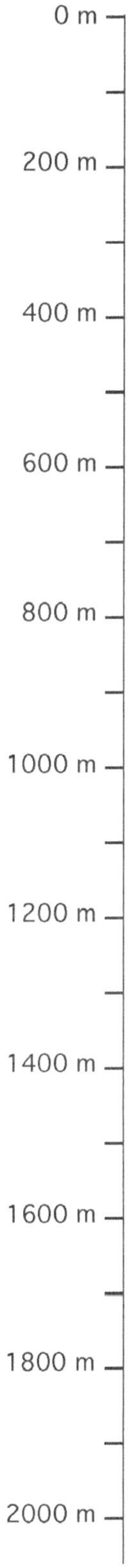

Figure 3 Scale for drawing a composite section.

Geological Exploration: The Geology of Gregwebb County, Texas: Part 2

Lab 10

Completing the Report

Your report indicates that Gregwebb County has good quality coal and warrants further attention. Using information provided by the cores and field mapping, evaluate the extent of coal deposits and the geologic structure of the area.

1. Record the fossil-determined ages for each sample site on the incomplete geologic map (Figure 4) and complete the job of drawing in geologic contacts on the map. Label the **age** and **lithology** of each outcrop belt.

   Note that there are two contacts on the map (the contact between points I and IV and the contact between points II and III) where boundaries of rock units occur in configurations unlike those in most areas on the map. These could be structural contacts or contacts of unconformity. Determine which type of contact is present in each case. For the contact between points II and III, determine why the trace of the contact is deflected in the upstream direction as it crosses the stream valley, **using the Rules for Interpreting Contacts on Geologic Maps below**.

2. Using the map, core data, and the strike and dip information on the map, construct and label geologic cross sections of the area from A to B and from C to D (Figure 5). Topography is provided on the blank cross-section diagrams. (Hint: It is easiest to do cross-section A to B first and then use it to help in constructing cross-section C to D.). Interpret the structure with the aid of diagrams showing examples of folding and faulting (Figure 6). **Read the section "Making a Geologic Cross Section" below.**

3. Use the two geologic cross sections, the completed map, and the lithologic log from core 3 to construct a subcrop map on the surface beneath the Cretaceous strata for the eastern part of the map area. (A subcrop map shows the geology that would be exposed on the ground if the overlying strata were stripped off.) A separate map sheet is provided to do this work (Figure 7). **Read the section "Making a Subcrop Map" below.**

4. Complete (and revise where necessary) the report on geologic history written after you compiled the composite stratigraphic section by adding a discussion on the structural history of the area. Check to make sure that your report summarizes the times of transgression and regression, times of marine and nonmarine deposition, times of uplift, erosion, and/or nondeposition (as determined from unconformities), and times and types of structural deformation.

5. Evaluate the map area for energy resources. Within the map area:

   A. indicate where coal is to be found at the ground surface, and

   B. indicate where coal is to be found in the subsurface.

   (Refer to the coal labeled on your cross sections, map, and subcrop map for guidance in this evaluation.)

Some Rules for Interpreting Contacts on Geologic Maps

A. Contacts between horizontal beds are parallel to topographic contours.
B. Contacts of beds that have dip angles lower than the stream gradient dip upstream.
C. Contacts of beds that have dip angles greater than the stream gradient dip downstream.
D. Contacts migrate downdip during erosion.
E. Contacts of vertical beds (planes) do not "V" or migrate position with erosion.
F. True dip angles can only be seen in cross section if the cross section is perpendicular to the strike of beds or plane of fault.
G. Anticlines and domes have their oldest beds in the center.
H. The axis of an anticline plunges toward the nose (closed end) of the structure.
I. Synclines have their youngest beds in the center.
J. The axis of a syncline plunges toward the open end of the structure.

Making a Geologic Cross Section

A. On a copy of the completed surface geologic map, fold the sheet along the A–B line.
B. Place the folded edge of the geologic map across the cross-section sheet, so the folded edge is aligned horizontal on the cross section and the boundaries of the map area coincide.
C. At each point where a rock unit contact crosses the folded edge, draw a perpendicular to the ground surface line on the cross-section sheet and mark the resulting point of intersection.
D. Using a protractor, draw in the dip of each bed on the cross section and project the rock unit contacts down below the ground surface.
E. Use the subcrop map to fill in the rock units underneath the flat-lying beds. Fold the subcrop map along the A–B line and follow the same procedures used with the surface geologic map.
F. Using the folded map, mark the points where the contacts between rock units cross the cross-section line at a stratigraphic level corresponding to the bottom of the flat-lying beds.
G. For rock units below the flat-lying beds, draw in the trend of each unit that intersects the subcrop surface by copying the dips determined on your subcrop map. Use a protractor to project the rock unit contacts below the base of the flat-lying beds on the cross section.
H. Check your cross section with your core correlation to see if they match. Make sure that all of the units present in the cores are found in your cross section.

Making a Subcrop Map

A subcrop map is one that is drawn over a surface that lies below the present ground surface, a map that shows the geology that would be exposed if the overlying strata or rocks were stripped off. It is drawn along surfaces such as unconformities that show the geology of the preexisting erosion surface. The intent is to show the appearance of geologic units on a surface other than the present ground surface.

Use the following steps to make a subcrop map for the geology of Gregwebb County.

A. Overlay a blank subcrop map sheet on your completed surface geologic map.
B. Mark and label the points where the rock unit contacts of the surface geologic map cross the left-hand edge of the subcrop map area.
C. Project structural boundaries (such as fault traces) beneath the overlying beds across an erosion surface produced during the early Cretaceous.
D. Project the contacts of formation units beneath the overlying beds, showing where you think they extend. If formation contacts would cross a structural boundary, decide how the formation boundary would change as it crosses the structure. At a fault, formation contacts would be offset. If it is a fold, the formation boundary would bend.

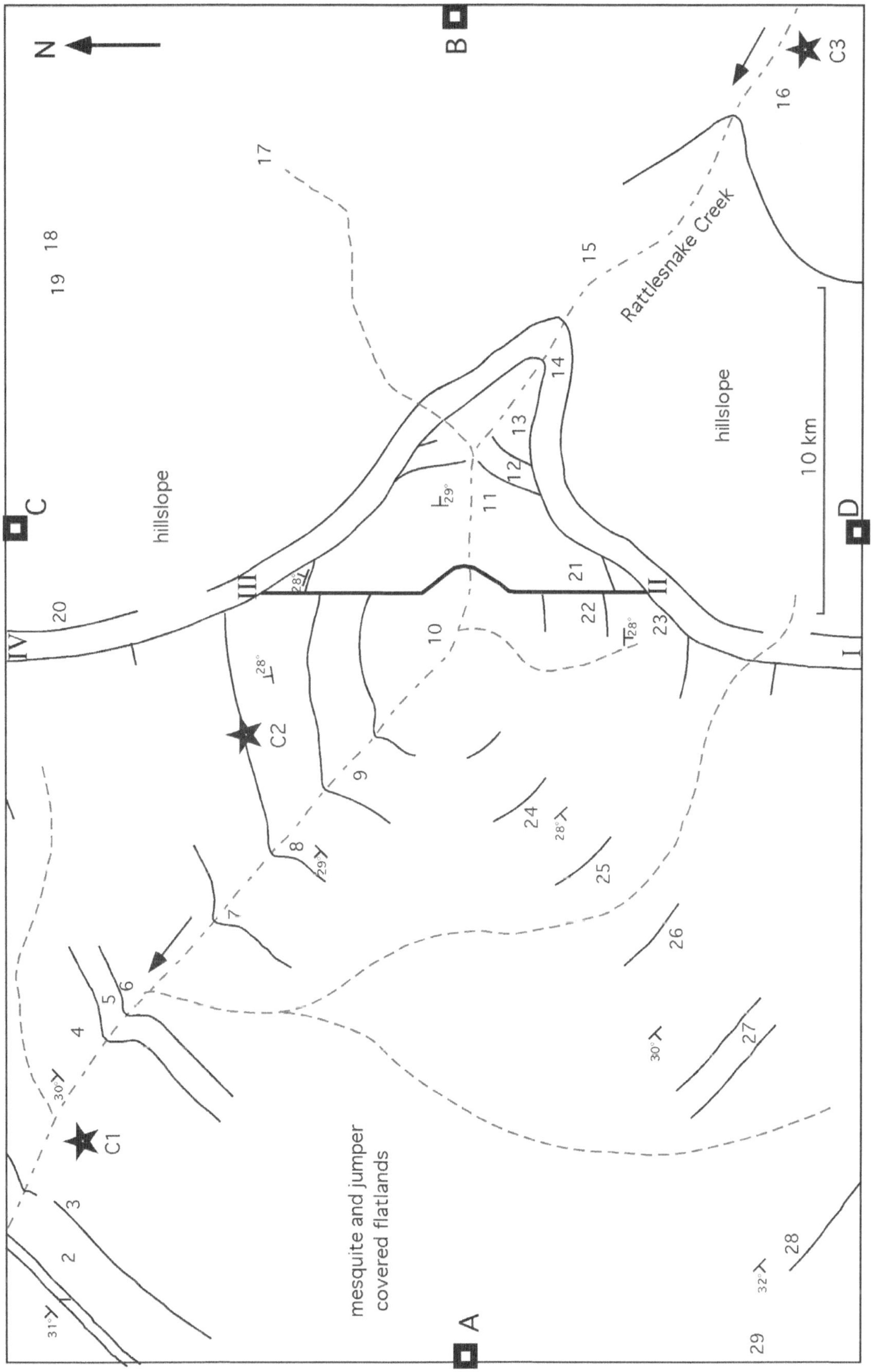

Figure 4 Incomplete geologic map for Rattlesnake Creek area.

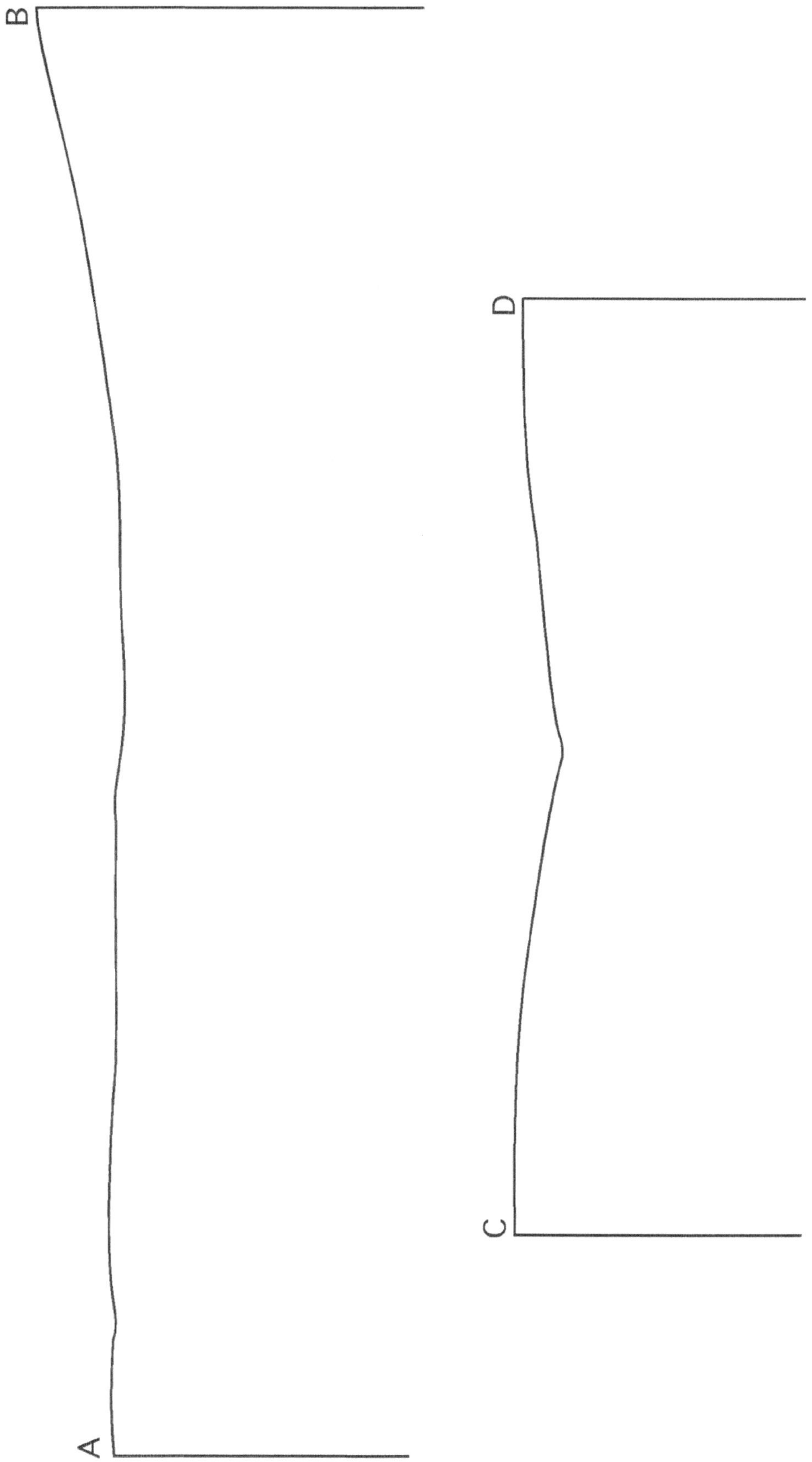

Figure 5 Ground surface profiles for vertical cross sections of transects A–B and C–D.

E. Project where you think the Early Carboniferous oolitic limestone formation extends beneath the overlying beds.
F. Fill in all of the subcrop map area with your projections of the rock units. To determine the position of formation contacts in the covered area, measure the width of unit outcrops on the west side of the map and assume the subcrops have similar width in the covered area. Use a ruler to plot their positions on the subcrop map.
G. Check to see that the rocks that occur in core 3 correspond with your subcrop map.
H. Use a new sheet of paper and make a clean, neat copy of the map. The subcrop map must be suitable to pass on to another geologist to work on. You are required to produce a geologically reasonable report and make it fit the data. Repeat until you have an acceptable version.

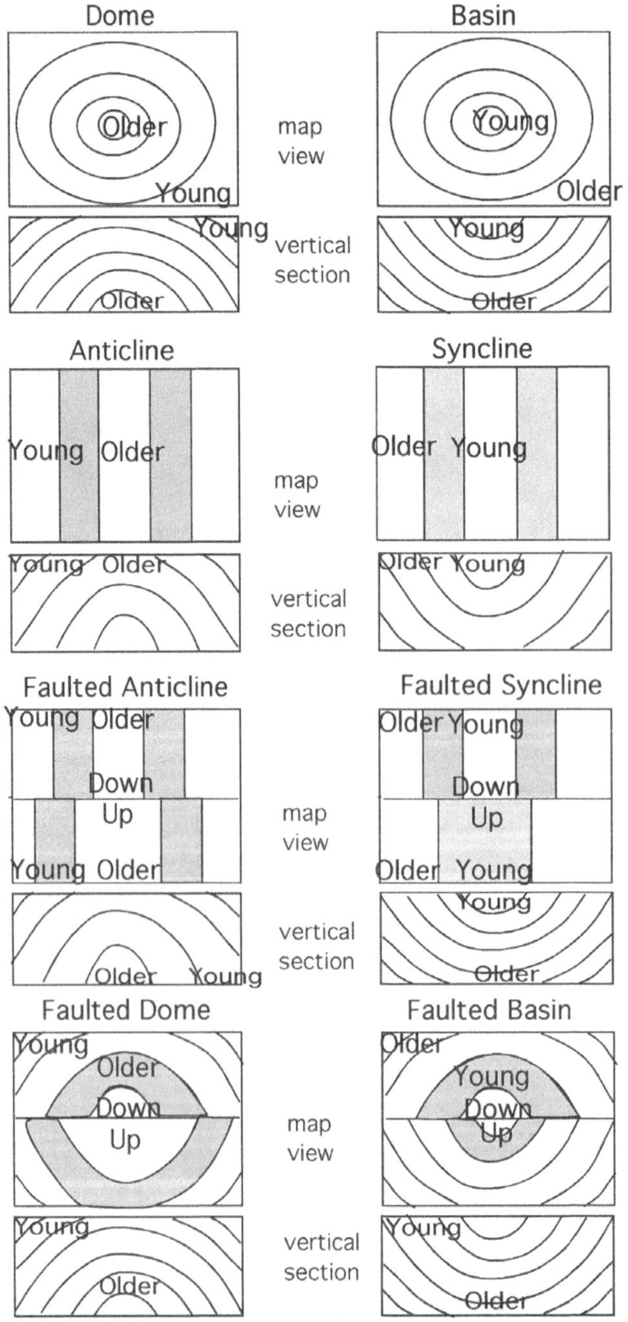

Figure 6 Examples of folded and faulted structures.

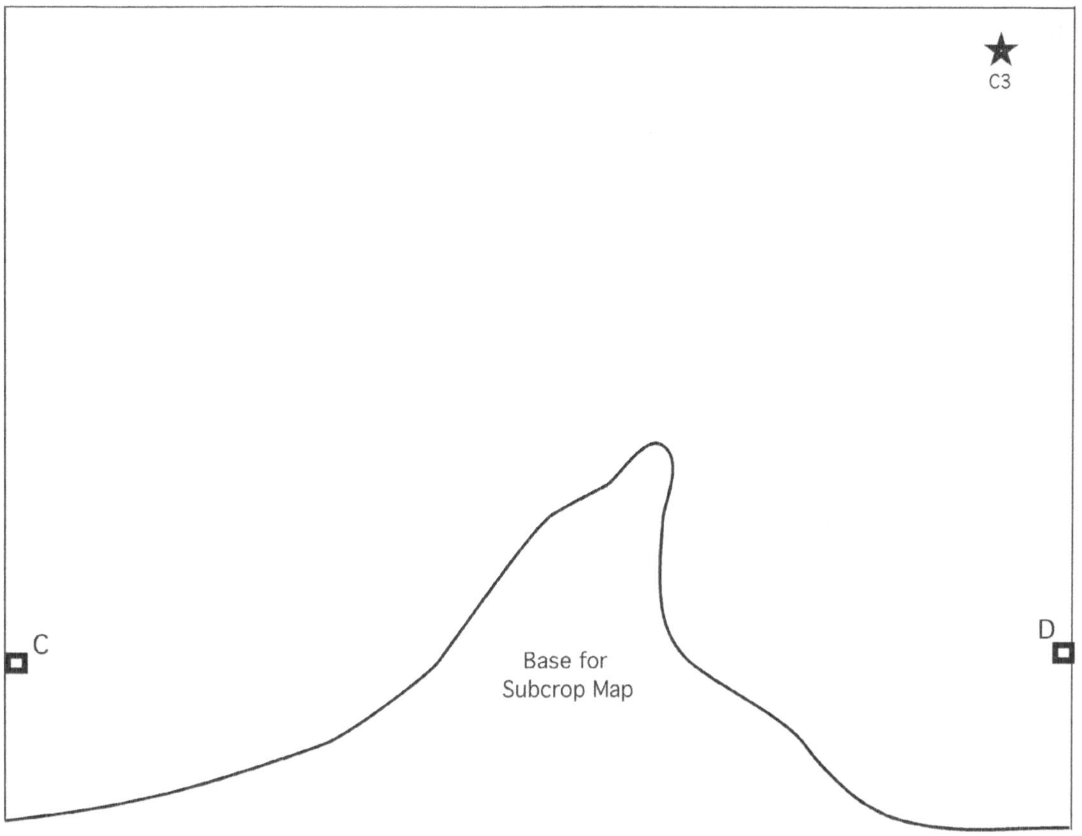

Figure 7 Base for subcrop map.

Questions

6. What type of geologic structures are present?

   A.

   B.

7. At what geologic time were these structures formed?

   A.

   B.

8. Is structural contact segments II and III limited to the area between point II and point III on the map or does it extend beyond this area? Why?

   (Be sure to show this on your subcrop map.)

9. A. What type of unconformity occurs beneath the conglomeratic Cretaceous sandstone?

   B. What type of unconformity occurs between the Cretaceous units and Eocene units?

10. Within the map area, where is coal to be found at the ground surface?

11. Within the map area, where is coal to be found in the subsurface?

    (Refer to the coal labeled on your cross sections, map, and subcrop map.)

Ocean Life and Trace Fossils

Lab 11

Marine Benthic Communities

The organisms living in a **biotic community** are linked by ecological interactions such as the flow of food and nutrients between species. Community members use food resources and space in ways that minimize competition yet maintains a flow of resources, a process called **niche partitioning**. Two aspects of niche partitioning are specialized **life position** (infaunal, epifaunal, nektic, etc.) and specialization to use **different food particle sizes** (minute particles, detritus and plankton, whole organisms.) When communities occupy different segments of an environmental gradient, they are arranged in **biofacies.** Controls on biofacies include substrate (rock, gravel, sand, mud), salinity gradient, depth gradient, and temperature. Placement within or below the **photic zone** is determined by noting the presence/absence of plants or organisms bearing symbiotic microalgae.

Life activities of some fossil species can be determined from their morphology or by predation damage. Coprolites and gastric residues provide an evidence of food consumed. Shell breakage or boreholes in shell are the evidence of predation by crabs or gastropods. Examine some fossils to determine life activity.

Inhabitants of the Seafloor (Common Fossil Producers)

Brachiopods (Phylum Brachiopoda)

Brachiopods were very common in Paleozoic oceans. Brachiopods are suspension feeders that use a fan of fleshy tentacles (the lophophore) inside the shell to catch very small floating food particles. They are living pumps that continually move water through the shell to trap food.

Summary of Characters for Brachiopods	**Age Range: Cambrian–Recent**
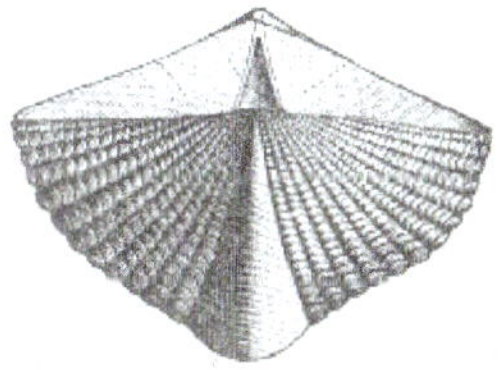 (Nicholson, 1879, Manual of Palaeontology, vol. 1, p. 440)	
Diagnostic Features	a two-part shell with a plane of symmetry passing through both valves; muscular pedicle stalk that attaches animal to substrate; lophophore support inside the shell; calcite or phosphatic shell
Morphology	muscular pedicle stalk that attaches animal to substrate; lophophore support inside the shell; calcite or phosphatic shell; plane of symmetry passing through both valves
Symmetry	mineralized shell with two valves
Life Habits	marine only; attached or semi-infaunal benthic suspension feeders

Brachiopods are distinguished from bivalves (clams) by having a bilateral plane of symmetry cutting through the center of both valves, so each side of the valve is a mirror image of the other side and the two valves are dissimilar from each other. Brachiopods are either **attached epifaunal** (fastened by a fleshy stalk) or **semi-infaunal** (living partly in sediment). Many species lost the pedicle as adults, living unattached. A globular form allows the shell to float on loose mud **(a boating effect)**, a large flat shell keeps a stable orientation on sediment, or growing numerous spines (productid brachiopods) provides support as **stilts or outriggers**.

Exercise Work

1. Look at the productid brachiopod specimens *Nudauris linospina* and *Grandaurispina vaga*. These productid brachiopods had numerous slender spines covering the valves during life and the concavo–convex shell is well adapted for semi-infaunal life on soft sediment.

   A. How did the spines help stabilize the shell in soft sediment?

   B. Is this shell morphology an adaptation for life in low-energy or high-energy environments?

2. The productid brachiopod *Hercosia uddeni* has abandoned the typical form of productid brachiopods and acquired different shell morphology.

   A. What group of non-brachiopod animals does the shell form of this brachiopod (*Hercosia*) most resemble?

   B. How did this brachiopod use its spines (those at the base of the cone; not the top—look at the clusters of shells) to position itself?

3. These dish-shaped brachiopods (*Rafinesquina*, #126) and spiny productid brachiopods (*Linoproductus*, #109; *Juresania*, #110) are preserved with both valves tight together. These brachiopods lived unattached on the seafloor and fed from water currents entering the shell margins. Which orientation would be best suited for stability and suspension feeding: with the concave valve facing up or down during life?

4. The large fold on the middle of the shell of these spirifer brachiopods (*Neospirifer*, #093; *Paraspirifer*, #099) serves to funnel water away from the shell after it has passed over the lophophore during feeding.

   A. If the fold is the exhaust zone, where does water enter into the shell?

   B. What is the manner of preservation of sample #099?

5. Brachiopods with an inflated globose form can be either semi-infaunal (floating on the mud) or epifaunal and attached to a hard surface by the fleshy pedicle. Epifaunal attachment requires strong pedicle support, which is indicated by a well-developed pedicle opening in the shell. A "floating" semi-infaunal habit utilizes low overall density (boating effect) and is associated with a very globose shell form, so the shell tends to have a small pedicle opening or none at all. Using the character of the pedicle opening to determine life habit, interpret the life habit (semi-infaunal or epifaunal attachment) of these inflated brachiopods (#104—*Composita*, #116—*Platystrophia*).

Specimen	Life Habit
#104	
#116	

Bivalves (Phylum Mollusca)

Bivalves (clams) are common in all aquatic environments, marine, and fresh water. Most bivalves are suspension feeders—living pumps that continually move water through the shell to trap protists and other plankton, although a few groups are deposit feeders or manage to culture microbes in their own tissue. Bivalves have a two-part shell with the two parts connected by a **ligament** made of tough organic material. The shell is distinguished by having a **plane of symmetry**, which passes between the valves, so one valve is a mirror image of the other. Some bivalves, such as oysters, secrete large, heavy shells composed mostly of calcite and are usually preserved whole, whereas most bivalves have a shell composed of aragonite.

Summary of Characters for Bivalves	**Age Range: Cambrian–Recent**
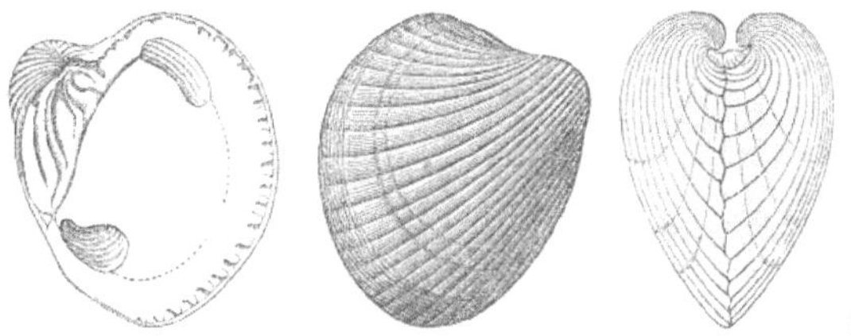	(Nicholson, 1879, Manual of Palaeontology, vol. 1, p. 440)
Diagnostic Features	valves connected by an organic ligament; mostly aragonite shell; plane of symmetry passing between the valves
Morphology	mineralized shell with two valves
Symmetry	bilateral
Life Habits	marine and freshwater aquatic; infaunal and epifaunal suspension feeders, some are cemented in place

Exercise Work

6. These are specimens of the bivalve *Venericardia* (#145, 371), a distinctive fossil of the Paleogene and one that is commonly found in local marine deposits (such as at Stone City Bluff on the Brazos River).

   A. Is this a burrowing infaunal or immobile epifaunal bivalve?

   B. What is the symmetry condition of the bivalve shell?

   C. Why are most bivalve fossils found as single shells rather than as paired shells joined together, like the brachiopod specimens mentioned in previous questions?

7. These are Cretaceous oysters (*Texigryphaea*, #186; *Exogyra* #168).

   A. Were they burrowing infaunal or immobile epifaunal bivalves?

   B. They have lost the characteristic bivalve symmetry. How can you tell them from brachiopods?

8. These large late Cretaceous oyster shells (*Exogyra*, #168, 176) had a life position with the big valve down.

   A. Suggest a function for the very thick shell that would be advantageous for the oyster. (At many times in their evolutionary history, oysters produced very thick shells.)

   B. Does the large size suggest abundant food supply or little available food?

9. These Cretaceous oyster shells (*Exogyra*, #169) show a history of shell destruction. There is **abrasion** on the shell, **borings** (by marine bivalves) into the shell, **fracturing** and breakage of the shell, and **sediment packed** into openings on the shell. Describe the sequence of events (a history) affecting this oyster shell.

   A.

   B.

   C.

   D.

   E. This *Exogyra* (#169A) has areas of shell that are rough and pitted and areas that are smoother. Smoother area is normal shell; pitted area is a result of boring into the shell. Draw a boundary line on the photo between the pitted and smoother and rougher parts of the shell. This is used to show exposed shell and shell covered by mud during life. Does the photo show a reasonable life orientation?

   F. The large groove on the thicker valve results from a growth interruption, when little shell was deposited.

   Suggest a reason for this type of growth interruption. (This type of feature can be seen on many clams and snails.)

10. Examine the outer surface of the oyster *Pycnodonte* (#184, 185). Note the attachment scar on each shell that shows an impression (xenomorphic sculpture) of the surface on which the oyster settled when it started growing. Identify one of the objects that served as an attachment surface for the young oyster.

Gastropods (Phylum Mollusca)

Gastropods (snails) are common in all environments on land and in the water, and some groups have even become planktic. Aquatic gastropods have gills for respiration, but those that have colonized the land surface have evolved a lung. Most gastropods are active animals, feeding by grazing, deposit feeding, or as predators, using a continuously growing strip of organic teeth (**radula**) in the mouth. The most distinctive feature of gastropods is the conical or coiled shell they secrete, which usually is coiled into a spiral form. It can be distinguished from nautiloids and ammonoids by the lack of internal septa. Many gastropods are active predators on other invertebrates.

Summary of Characters for Gastropods	Age Range: Cambrian–Recent
	Woodring, W. P., Bramlette, M. N.& Kew, W. S. W., "Geology and paleontology of Palos Verdes hills, California," *U.S. Geological Survey professional paper, 207*, Washington, U.S. Government Printing Office, pl. 35, 1946.
Diagnostic Features	torsion of soft organs of body
Morphology	tubular shell lacking regular chambers, which is normally coiled in a spiral form, ranging from helicospiral to planispiral
Symmetry	bilateral (of soft body); shell is varied
Life Habits	vagrant benthic grazers or predators, some are planktic

Exercise Work

11. Species of cone snails are carnivorous and catch prey by jabbing them with poison. Species feed on different types of prey—worms, other mollusks, or fish. The size of the aperture (shell opening) correlates with the type of prey: **fish eaters** (piscivores) have large apertures, **worm eaters** (vermivores) have a narrow aperture, and **mollusc eaters** (molluscivores) have intermediate size apertures. These features are easily seen on fossil shells and make determination of diet possible.

    A. These cone shells (*Conus*, #192) preyed upon worms and burrowed through sediment to find them.
    If the shell is streamlined for burrowing, which end (narrow or wide) of the shell would be in front during movement?

    B. Compare the shells of *Conus* (#192) with shells of *Turritella* (#206). Which one (#192 or #206) is better adapted for burrowing through sediment?

12. **Predator–prey interactions**. This gastropod (*Pleuroplectella*, #199) shows breakage typical of crab or lobster predation and subsequent repair. Predators may not be successful in breaking thick shells or shells with narrow apertures (shell openings).

    How many unsuccessful predation attacks (recorded by arcuate breakage and subsequent shell repair) can you see on the shell?

13. **Predator–prey interactions**. Mollusks are used for food by many predatory animals. These Eocene surface-dwelling turritellid gastropods (*Mesalia*; #414, #415, and #416) were attacked by crabs (note the shell breakage) or drilled by predatory gastropods (note the holes bored through the shell). Indicate which samples show **predation by crabs, predation by gastropods**, or **non-lethal injuries**? (Look at #199 to see examples of crab predation).

Specimen	Type of Predation Activity
#414	
#415	
#416	

14. **Species interactions**. These specimens are of a coral (*Septastrea*, #425) growing over a gastropod shell; but, the coral has not completely overgrown the shell. There is an opening (aperture) in the coral cover, giving access to seawater for an animal living in the shell. But, the apertural opening is not made of gastropod shell, it is formed by coral skeleton and we can infer that the shell was occupied by a hermit crab after the death of the gastropod and the coral continued to grow on the shell. The coral could have settled on the gastropod shell when the gastropod was alive, forming a coral–gastropod association, or after the gastropod died and the shell was occupied by a hermit crab, forming a coral–hermit crab association.

    What is more probable: a coral–gastropod association or a coral–hermit crab association? Why?

15. **Predator–prey interactions**. These are cone shells, predatory gastropods that attack prey animals by using harpoon-shaped teeth to stab and inject poison to kill it. Cone shell poison is very toxic and some cone shells are known **man killers**, in which shell collectors picking up large cones (especially, *Conus geographus*) are stung by it and die from the poison. Species of cones are specialized to feed on different types of prey—worms, other mollusks, or fish. The width of the aperture (shell opening) correlates with the type of prey: **fish eaters** (piscivores) have wide apertures, **worm eaters** (vermivores) have a narrow aperture, and **mollusk eaters** (molluscivores) have intermediate width apertures.

    A. What is the diet of the species *Conus generalus* (#379), *C. geographus* (#380), *Conus striatus* (#382), and *Conus textile* (#381)?

Specimen	Diet	Specimen	Diet
#379		#380	
#382		#381	

    B. What feature suggests that *C. geographus* is a very dangerous cone to handle when alive?

Crustaceans (Phylum Arthropoda)

Crustaceans (crabs, lobsters, shrimps, etc.) are the most varied and successful of the aquatic arthropods in modern seas, displacing trilobites during the latest Paleozoic. Being arthropods, crustaceans have a segmented body with a stiff exoskeleton bearing appendages and mouth parts and molt many times during life. Molting results in fragmenting the exoskeleton, so complete body fossils usually are the result of death, especially if the crustacean digs down into the sediment for protection.

Summary of Characters for Crustaceans	Age Range: Cambrian–Recent
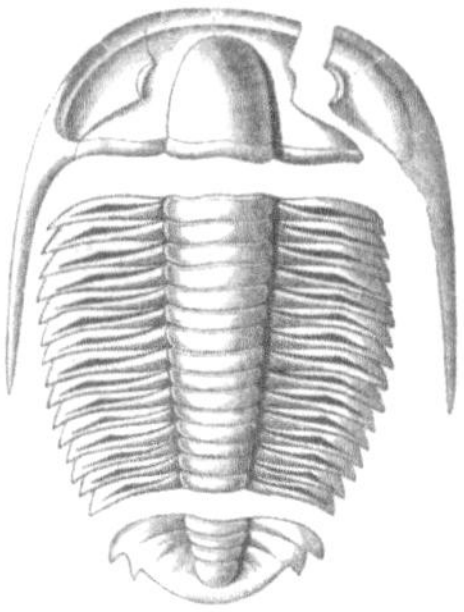 (Nicholson, 1879, Manual of Palaeontology, vol. 1, p. 352)	
Diagnostic Features	have five pairs of head appendages

Summary of Characters for Crustaceans	Age Range: Cambrian–Recent
Morphology	body with head–trunk portion and segmented tail and segmented appendages plus antennae; weakly to moderately mineralized exoskeleton
Symmetry	bilateral
Life Habits	mobile aquatic nekto-benthic predators

Echinoderms (Phylum Echinodermata)

Echinoderms (starfish, sand dollars, sea urchins, crinoids, sea cucumbers, etc.) are an unusual group of animals living in the oceans. They have a **plated skeleton** formed of single crystal plates of calcite secreted just below the skin of the animal. The plated skeleton acts like armor. They also have a hydraulic system (the **water vascular system,**) that uses sea water as hydraulic fluid to move body parts. The other major feature is their **pentameral (five parts) symmetry**, with a less noticeable bilateral symmetry. The skeleton plates of echinoderms are held together by tissue that disintegrates after death, releasing large amounts of bioclast debris on the seafloor.

Echinoderms have diversified in morphology more than most phyla. They are either epifaunal suspension/detritus feeders (crinoids), slow moving herbivores (echinoids), slow moving deposit feeders (sea cucumbers), or slow moving predators (starfish). The common echinoderms of the Paleozoic were crinoids—stalked animals with feeding arms. The common echinoderms of the Cretaceous and Cenozoic are sea urchins and sand dollars—free-living animals (both epifaunal and infaunal burrowers) covered with spines.

Summary of Characters for Echinoderms	Age Range: Cambrian–Recent
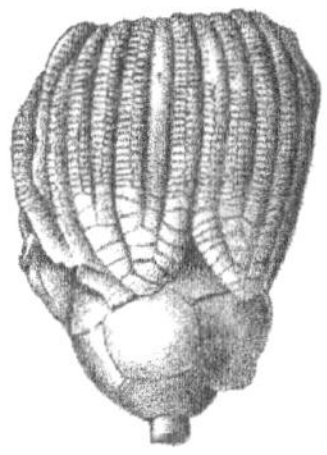 (Contrib. to Paleontology, C. White, 1880, U. S. Geol. Survey., pl. 40)	
Diagnostic Features	internal plated skeleton composed of single crystal calcite plates; body with water vascular system; pentameral symmetry
Morphology	varied; have a plated skeleton
Symmetry	pentameral
Life Habits	marine only; attached suspension feeders, mobile benthic detritus feeders

Echinoids

Echinoids are bottom dwellers species with **long, stout spine, and large spine bases** are surface dwellers, while species with **short-slender spines and small spine bases** live buried in sediment. Echinoids with smooth domed tops (non-grooved) or flattened ovoid shapes live in sandy sediment; species with a high slanting surface and grooved top live in mud sediments. The much flattened shape of sand dollar echinoids is an adaptation to life on a sandy substrate.

Sand Dollars

These distinctive and easy to recognize echinoids have a flattened disc form, short spines, and a flower-like pattern on the upper surface. They have adapted to life on shifting sands in a high-energy, shallow–marine environment. They appear in the Eocene, so are restricted to Cenozoic strata. See discussion in Functional Morphology.

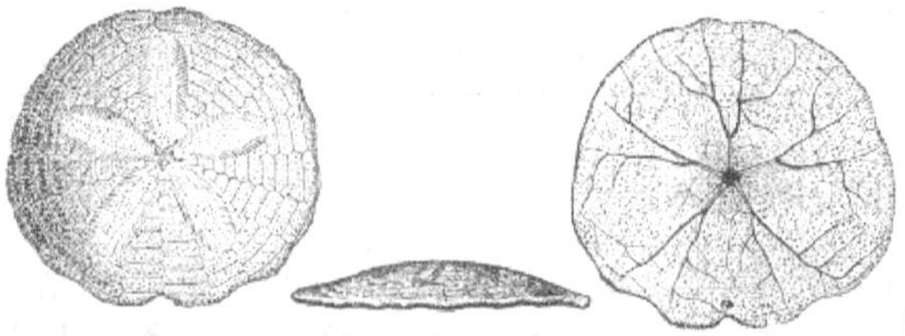

(Nicholson, 1879, *Manual of Palaeontology*, vol. 1, p. 228)

Figure 1 Sand dollars, top, bottom and side views.

Exercise Work

16. This a crinoid (*Rhodocrinites*; #273 and #282), an animal that has the common name "sea lily," and looks somewhat like a plant.

    A. How can you tell this is an echinoderm and not a plant?

17. *Asteriacites* (#409). These trace fossils were formed when an animal left an imprint of its body in the sediment, producing a depression that was subsequently filled with sand. The traces therefore are preserved on the undersides of the slabs. What type of echinoderm animal made this imprint?

18. These echinoids (*Holectypus*, #298; *Holaster*, #299) are common in Cretaceous strata of Texas and had tiny spines covering the shell, in contrast with the large spines present on *Cidaris* (#296).

    A. Since the spines are missing from the fossil, how can you tell that the spines were much smaller than those present on *Cidaris* (#296)?

    B. Were these echinoids infaunal or epifaunal? (See discussion of form and function of echinoids.)

    C. Were these likely to be predators, attacking and eating animals, or herbivores–detritivores?

19. These are sand dollars (*Dendraster*, #286; *Mellitella*, #413), flattened echinoids that live partly buried in sand in nearshore settings. Sand dollars are good indicators of shallow-water environments.

    Is the flat disc shape of sand dollars an adaptation to stability in higher energy conditions of waves and tidal currents? (Answer yes or no)

20. Determine the sediment grain size in which these infaunal Eocene echinoids (*Eupatagus*, #302) lived. Infaunal echinoids with smooth (non-grooved) domed tops and large spine bases live in sandy sediment while infaunal echinoids with a high slanting surface and grooved top and small spine bases live in mud sediments.

    A. Is the top domed and smoothed or deeply grooved?

    B. Are the spine bases small or large, relative to those on *Holectypus* (#298) and *Holaster* (#299)?

    C. Did this species live in muddy sediment or sandy sediment?

Trace Fossils

In contrast to skeleton remains **(body fossils)** of a once-living organism, **trace fossils** are the marks in sediments produced by the activities of organisms. **Trace fossils show activities of walking, burrowing, feeding, sheltering, rooting, constructions (nests, etc.), excrement (coprolites), rocks swallowed to grind up food (gastroliths), and sediment layers trapped by organisms (stromatolites)**. An organism can make many types of traces by its different activities.

Types of Trace Fossils

Locomotion traces—tracks and trails, usually on surface of sediment.

Resting traces—markings showing outline of animals, made during rest.

Feeding traces—often complex patterns of feeding activities.

Shelter burrows—burrows (usually simple) used for long-term shelter.

Borings—permanent shelter burrows excavated into hard rock or shell.

Coprolites—fossil excrement (feeding).

Rooting traces—impressions of roots of attached organisms.

Stromatolites—layers of sediments shaped by microbial mats.

Exercise Work

Specimens	Type of Trace
21. *Skolithos* (#305) Vertical traces are common in shallow-water environments, where animals dig deep to avoid being washed away during high-energy conditions (waves, storms).	
22. *Ophiomorpha* (#033) and *Gyrolithes* (#372) are tunnels made by burrowing shrimp. The lumpy walls of *Ohiomorpha* are made by placing pellets along the walls of the tunnel. *Gyrolithes* is a vertical spiral entranceway into a complex burrow system, the complex form acting as a baffle to deter predators entering the burrow.	
23. *Cruziana* (#347 and #431) and *Rusophycus* (#417) are the trackways and resting traces made by trilobites. They are common in early Paleozoic strata, when trilobites were common and large.	
24. Trackways (#369) are trackways made by arthropods, but note the differences in number of legs and direction of leg motion relative to direction of movement when compared with *Cruziana* (#347) trackways.	Were these trackways (#369) made by trilobites?

Specimens	Type of Trace
25. *Nereites* (#427). This trace was formed by the repetitive behavior of an organism systematically moving sediment. This type of trace is common in deep-water environments.	
26. Reptile tracks (#340) often occur in association with mudcracked sediments.	Depositional environment?
27. Coprolites (#348) is fossil excrement. What might it tell you about the feeding habits of the organism?	Type of organism?

Reefs and Reef Builders

Lab 12

The Reef

A **reef** is both an elevated accumulation of sediment and shell that is formed of **close-packed calcareous skeletons in a framework** and can **grow upward to large size**, rising above the surrounding seafloor (has relief). Reefs occur in marine waters of low productivity (contain low levels of nutrients), but support diverse organisms and high biomass because many reef organisms house algal symbionts living within their tissue that provide them with food. Modern reefs are common in waters within **25° North and South latitude** from the equator and **17°C–30°C**. Reefs do not occur in high nutrient waters because the symbiotic algae need light to live. Reef communities produce an "island" of high productivity and diversity where nutrients are quickly cycled among resident species.

Reef Zonation. Reefs contain multiple environments occupied by different communities of species and produce different types of sediment. On an atoll or fringing reef, the typical succession of reef zones between land and open ocean is: **lagoon**, **reef core**, **reef margin**, and **fore reef slope**. Reef growth on shelf margins can create a raised rim on the continental shelf enclosing a deeper lagoon. Reef trends can be predicted with facies models and can be imaged with seismic data.

Reef Types. The familiar reef is a carbonate structure that corals build up to sea level by growing one on another and producing a solid framework that resists waves and storms. This produces a **barrier reef** with a steep fore reef front. Other areas of carbonate sediment with reef structures have small **patch reefs or mounds** that grow above the surrounding seafloor, but do not grow to sea level. They are common on shelf surfaces with a low angle of slope. This type of carbonate shelf is called a **carbonate ramp**.

Ecological Roles of Reef Organisms

Organisms occupy several ecological roles on the reef. These include the following:

Frame builders—organisms that grow into a frame structure and remain attached after death.

Binders—organisms that grow as encrusting layers over sediment and frame (like poured concrete).

Bafflers—organisms that trap sediment and hold it without cementation, such as patches of sea grass or sponges or branching coral

Sediment producers—organisms whose skeletons quickly break into pieces, producing skeletal bioclasts; many organisms (especially algae) produce small bioclasts that create most reef sediment.

Bioeroders—organisms whose activities cause fragmentation of skeletons, producing skeletal debris.

Exercise Work

1. The Cretaceous Stuart City reef trend of central and coastal Texas is an important reef complex in the state. Below is a diagram (Figure 1) showing reefs in this reef trend.

   A. How much topographic/bathymetric relief occurs on the reefs on this shelf?

   B. What type of reef deposit is present in this reef trend? (barrier or patch)

   C. What type of slope occurs seaward of the beach zone (steep slope or ramp)?

   D. What are the horizontal and vertical dimensions of the reef trend?

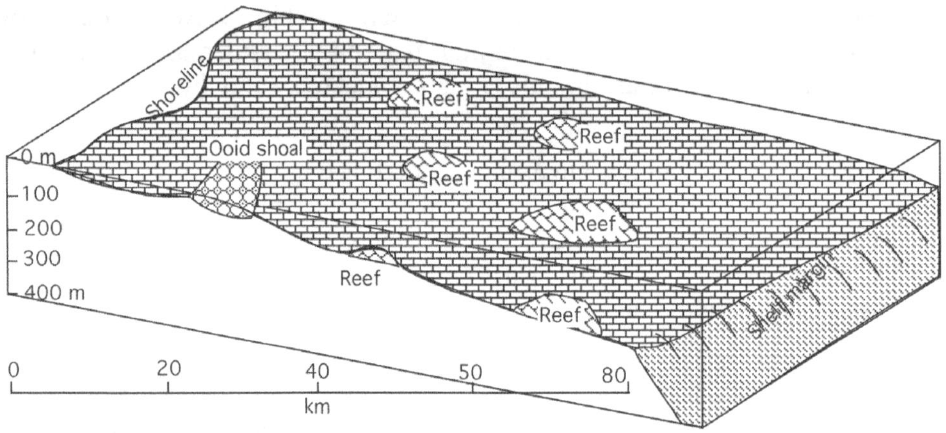

Figure 1 Cretaceous reef deposits of Texas.

2. The Permian Capitan reef trend of west Texas, part of which is well exposed in the Guadalupe Mts. of west Texas and adjacent New Mexico, is the other great reef complex that is an exploration target. Below is a diagram (Figure 2) showing a cross section of the Permian Capitan reef trend in the Guadalupe Mts.

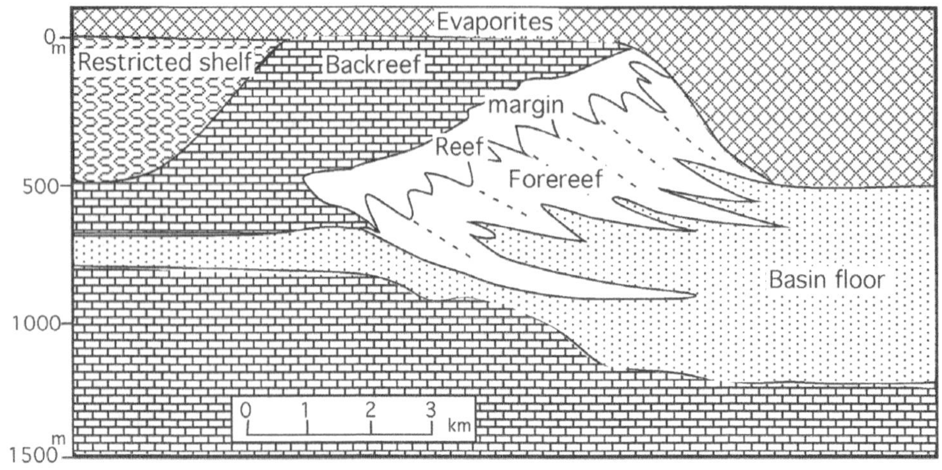

Figure 2 Capitan reef cross section of the Delaware basin.

A. How much topographic/bathymetric relief occurs on the reef margin in the Guadalupe Mts?

B. What type of slope occurs seaward of the reef (steep slope or ramp)?

C. What type of reef deposit is present in the Capitan? (barrier or patch)?

D. What are the horizontal and vertical dimensions of the reef trend?

E. Of the two examples of Texas reef trends (Cretaceous or Permian), which is most similar to that of modern reefs?

Reef Builders

Colonial organization and modules. Most reef builders are colonial organisms with many units (zooids) that are interconnected by living tissue. This is true of most **reef corals** (Phylum Cnidaria) and **all bryozoans** (Phylum Ectoprocta). Bryozoans and many **sponges** (Phylum Porifera) have modular organization with units controlling the flow of water currents bringing fresh water to feeding areas. When colonies with modular design are broken (by storms, etc.) the fragments can grow separately into genetically identical clones of the original colony. This gives clonal organisms an asexual as well as sexual means of reproduction and an indefinite life span.

Corals (Phylum Cnidaria)

This phylum includes reef-building corals, the dominant frame builders, and sediment producers of modern reefs. Corals are restricted to marine environments and are epifaunal, benthic micro-carnivores that kill their prey with stinging cells on the tentacles. Most reef corals are colonial. Corallites contain **septa** (partitions running lengthwise along the inside of the tube) and **tabulae** (partitions that divide the tube into chambers).

There are three main groups of corals (Figure 3) in the fossil record, two of which went extinct during the end-Permian mass extinction event. Corals forming framework of reefs today are the scleractinian corals. During the Paleozoic, solitary rugosan corals were most common and often occur as great piles of corals, forming patch reefs in the Devonian. They are common in marine deposits from the Ordovician to Permian. Scleractinian corals are descendents of rugosan corals.

Summary of Characters for Rugosans and Scleractinians	**Age Range (Rugosa): Ordovician–Permian** **Age Range (Scleractinia): Triassic–Recent**
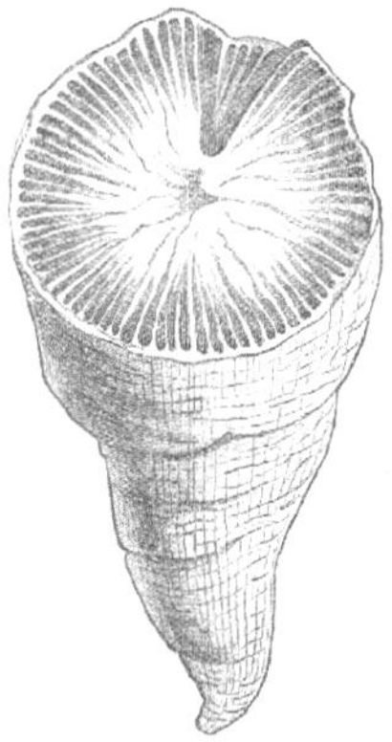 Rugose coral (Nicholson, 1879, Manual of Palaeontology, vol. 1, p. 209)	
Diagnostic Features:	radial pattern of septa in corallite; skeleton made of calcite (rugosa) or aragonite (scleractinia)
Growth Forms:	horn shaped (solitary); colonial, with massive or branching colonies
Life Habits:	live attached to seafloor or hard substrate

Bryozoans (Phylum Ectoprocta)

This inconspicuous group has produced an abundance of fossil material. They produce a carbonate skeleton that looks like a miniature coral colony, except that the tubes do not contain vertical septal partitions. The zooids share food and other resources, allowing some to perform specialized functions without having to feed for themselves. Bryozoans are mostly sediment producers or encrusters and rarely dominant in reefs.

A distinctive group of bryozoans are the **fenestellids**, (Figure 4) that produced a colony resembling a lace fan. During the Carboniferous, fenestellids grew in a spiral colony form, with the fenestrate sheet attached to a screw-like central rod known by the common name *Archimedes* screw (Figure 4), hence the formal name *Archimedes*. *Archimedes* is common in early Carboniferous strata. Fenestellids were very common in patch reefs and were able to trap sediment in the reef.

Summary of Characters for Fenestellid Bryozoans (Phylum Ectoprocta)	**Age Range Ordovician–Recent**

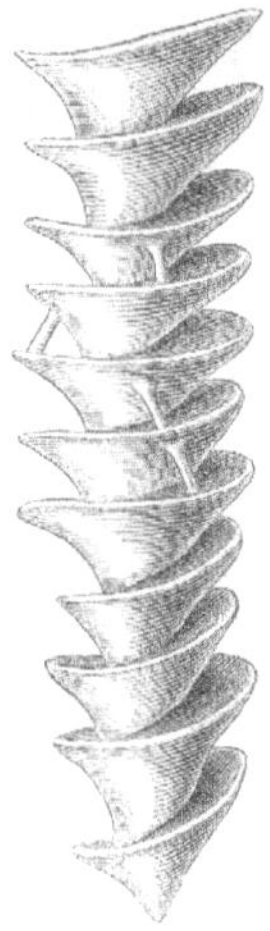

Archimedes, a fenestellid bryozoan

(Nicholson, 1879, Manual of Palaeontology, vol. 1, p. 423)

Diagnostic Features:	colony of tiny tubes or chambers, much smaller than corallite tubes, with specialized reproductive chambers
Growth Forms:	massive, encrusting, and fenestrate
Life Habits:	benthic suspension feeders; live attached to seafloor or hard substrate

Other Reef Builders

Calcareous Algae. These are common frame-building components of modern reefs. Because they are algae, they are confined to the photic zone and fill a role similar to that of corals on the outer, wave-pounded margins of reefs. Calcareous algae are prolific sediment producers in many tropical marine environments. In deeper waters, they grow in the form of balls or lumps called **rhodoliths**. This growth form also occurs at the foot of reefs where growth rates are slow.

Stromatoporoid sponges. Stromatoporoids were a reef frame builder during the mid-Paleozoic and early Mesozoic as coral-head masses with variable form and internal layering, but are uncommon at other times. Stromatoporoids produce a massive basal carbonate platform (in addition to the normal spicular sponge skeleton).

Stromatolites. (Cyanobacteria). In many reefs, microbial mats are a producer of cemented reef rock. These **microbial mats** do not grow with the simple, laminated character seen in common stromatolites, but produce highly irregular surfaces and encrusting sheets that bind other reef components by precipitating carbonate within the sediments. In the absence of true skeleton producers, stromatolites can produce reefs, such as the very large Precambrian stromatolite reefs. Distinctive features include microscopic laminated or clotted texture.

Exercise Work

3. This is a piece of Cretaceous reef rock (#57) from reefs fringing the central Texas high. The shells in this sample have all been replaced with silica.

   A. Is the dominant organism of this reef a coral? (Answer yes or no)

   B. Is this a high- or low-diversity fossil assemblage?

   C. Is the dominant organism the same as the dominant group in modern tropical reefs? (Answer yes or no)

   D. What is the carbonate rock type name for this rock? Although it is now silica, it formed as a carbonate sediment. (Use the carbonate rock classification: micrite, wackestone/packstone, grainstone, boundstone.)

4. This is a piece of reef rock (#R101) from a patch reef formed by algal mats. The algae produced cement that bound the sediment (see Stromatolite paragraph above). The rock has a clotted, irregular fabric and contains some small particles.

   A. What produced the framework for this reef deposit?

   B. What is the abundance of shells in the rock? (rare, few, many, or abundant)

   C. What is the carbonate rock type name for this rock? (Use the carbonate rock classification: micrite, wackestone/packstone, grainstone, boundstone.)

5. This is a rock (#A9) with abundant fossils in a fine-grained matrix. It is called Texas rock and is used for decorative exterior surfaces.

   A. Is this a reef sediment?

   B. Is the dominant large fossil able to grow onto other shells and bind the sediment to form a reef?

   C. Suggest a reason why the shells are dissolved but the carbonate matrix is still present.

6. This colonial branching coral, *Acropora* (#065), is common on modern reefs. Each small opening in the branch contained a coral animal and there is much skeletal material between the corallites, secreted by common tissue connecting the coral animals.

   A. Is this colony morphology a type that is easily fragmented by storms? (Answer yes or no)

   B. Compare this specimen with the Eocene coral *Acrohelia* (#072). Are they similar in general form?

   C. Would Eocene *Acrohelia* have the ability to grow into a new clone colony if a branch was broken from the parent colony?

7. The holes in this scleractinian reef-building coral (#067) were made by boring clams (bivalves), an important group of bioeroders. Bioeroders on modern reefs include sponges, clams, sea urchins, and fish. Parrot fish bite off pieces of coral to feed on coral tissue and microalgae and then excrete the coral fragments. Bioeroders are major producers of loose sediment on reefs. What effect do these bioeroders have on the reef framework?

8. These (#037) are specimens of a bryozoan with a branching growth habit. The colony surface has distinctive star-like raised areas that correspond to integrated module units in the colony and aided the flow of water over and away from the surface of the colony.

   A. Would this bryozoan be a frame builder, sediment binder, or sediment producer on a reef?

   B. If this colony were broken into pieces, could the pieces continue to grow to form a new colony?

   C. Why would a colonial organism organize its zooids into modular units? What is the advantage?

9. These are fenestellid bryozoans (#039), named for the fenestrae (Latin for windows) that occur in the lattice-work portions of the colony. Most fenestellid colonies grew as simple fans or cup-shaped colonies. Some fenestellid colonies have a central portion that twisted as it grew, producing a spiral axis of thickened skeletal material. This is called *Archimedes* (#027, 374) and is commonly found alone, stripped of the more delicate fenestrate portions of the colony.

   A. Would you consider fenestellids frame builders, binders, or sediment producers on a reef? Give reasons for your answer.

   B. Would you expect fenestellids to live in a high- or low-energy environment?

10. These slabs show specimens of blade-like calcareous algae (#350 and #351), late-Paleozoic algae that grew as wavy sheets (typically seen as "potato chip" pieces) on the seafloor. They were the dominant component of patch reef banks of Carboniferous age, often to the exclusion of animal fossils.

    A. What is the ecological role of these algae in Carboniferous reefs? (frame builder, binder, sediment producer, or bioeroder) Do you see evidence of encrusting growth?

    B. Did they form a solid framework?

    C. Did these algae live in a low- or high-energy environment?

    D. Name the rock type (carbonate rock classification) of this rock sample. (Use the carbonate rock classification: micrite, wackestone/packstone, grainstone, boundstone.)

11. Examine stromatolites #342 and #357.

    A. Judging from the grain size of the light-colored matrix between the stromatolite domes of #342, what energy levels were present in this depositional environment?

    B. How do the stromatolites numbered #357 differ from the #342 stromatolites?

    C. What is the rock type (rock name) of the sediments containing the stromatolite structures? (Use the carbonate rock classification: micrite, wackestone/packstone, grainstone, boundstone.)

12. This specimen (F20) is Precambrian in age. Which reef builder described above made this structure?

Paleoclimates and Ancient Oceans

Lab 13

Controls on Climate

Climate refers to the physical conditions of the atmosphere and the oceans on Earth and the variations in these systems through time. The average climate of any location is ultimately determined by the amount of **solar heat energy** received at that place, which varies geographically along the curvature of the Earth and seasonally on a yearly cycle. **Currents** in the atmosphere (Figure 1) and oceans **redistribute the heat energy** over the Earth. Circulation patterns in the oceans and atmosphere are closely linked, although the patterns are modified by the presence of land masses that form barriers to flow. The aspects of climate most apparent to life are average **temperature** and annual **rainfall** in an area. These control the distribution of organisms, which adapt to rainfall condition and temperature regimes in ways that are identifiable in modern and ancient biotas, making climate reconstruction possible. While the amount of solar heat received by the Earth is nearly constant, variations in composition of the atmosphere and reflectivity of the surface of Earth (**albedo**) control the ratio of heat absorbed to heat lost, producing long-term change in the global heat budget.

Rainfall controls: Global patterns of rainfall are determined mostly by the **major circulation cells of the atmosphere** (Hadley, Ferrel, and Polar cells). These occur in latitude-parallel bands, producing belts of high rainfall and low rainfall (Figure 1). Latitudinal zones of rising air produce high rainfall because air holds less water vapor as it cools. Zones where air descends produce dry areas because descending air can hold more water vapor as it warms, drying out the land surface. **Secondary control** comes from interaction of ocean currents with trade winds and the westerlies. This interaction often produces dry areas on the west sides of continents and higher rainfall on the east sides of continent. Regional patterns of rainfall are modified by surface elevations on continents (Figure 2).

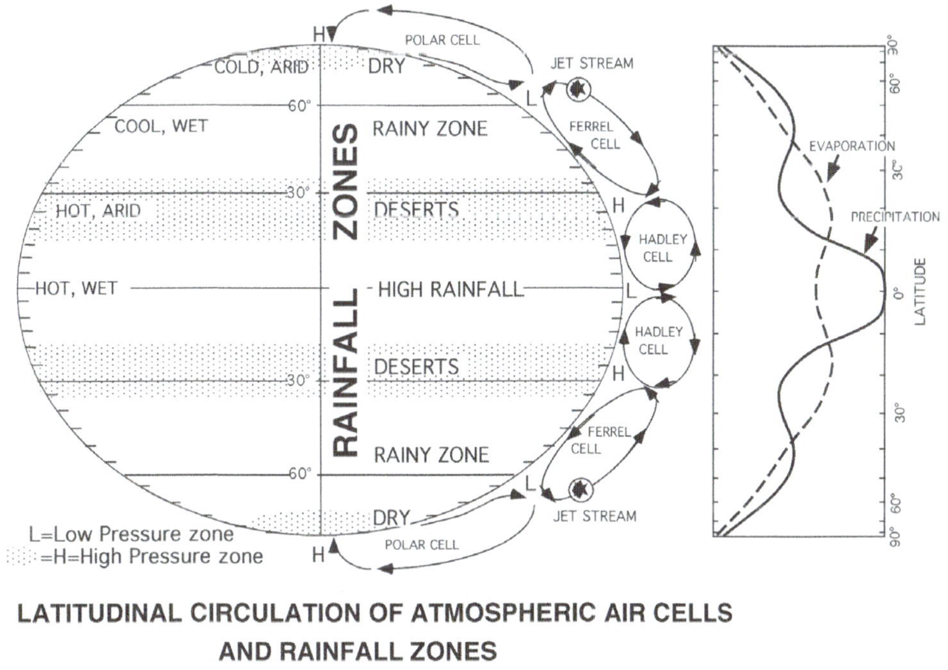

Figure 1 Latitudinal rainfall zones and wind directions and circulation of atmospheric air cells.

Latitudinal Climate Zones

The combination of temperature gradient and atmosphere circulation cells produces a basic set of four latitudinal-parallel climate zones (Figure 2) in each hemisphere of the Earth. Heating in the tropics causes air to warm and rise in the atmosphere, generating rainfall and creating the warm, rainy **equatorial climate zone**. Cooled, dried air descends to the surface in the latitudes of the Tropic of Cancer and Tropic of Capricorn, forming broad, stable areas of higher air pressure over the **low-latitude desert zones**. These zones are paralleled by a **cool temperate rainy zone** in mid-high latitudes. The polar regions are sites where cold, dry air descends to the Earth, producing **polar arid regions. These remain ice covered** because of low evaporation rates. The margins of the climate zones are modified by ocean and surface air currents, which must be plotted in models to determine detailed climate patterns.

Circulation Systems

The movement of air and water occurs in layers, but variations in topography and bathymetry interferes with regular patterns of flow and controls heat distribution and rainfall distribution.

Oceanic circulation: There are two major circulation systems in the oceans; a **latitudinally controlled** surface current system and a **density-driven** deep ocean current system. The surface current system consists largely of latitudinal (E-W) ocean current cells, which are driven by the surface winds (Figure 3). When these currents come to the edge of a land mass, they force cool, nutrient-rich water to the surface in upwelling zones. Upwelling zones create cool climates where they occur.

Regional controls on climate

The latitudinal climate zones produced by primary temperature gradient and atmospheric circulation cells are frequently modified by secondary factors, producing distinctive regional climates. These can be predicted in climate and circulation models.

East coast rains and west coast deserts: The low-latitude trade winds bring **ocean moisture** over the eastern sides of continents and pull **dry air** over the western sides of continents. This produces large areas of wet climate on the eastern sides of low-latitude continents and an east–west rainfall gradient (as seen in North America and on southern hemisphere continents).

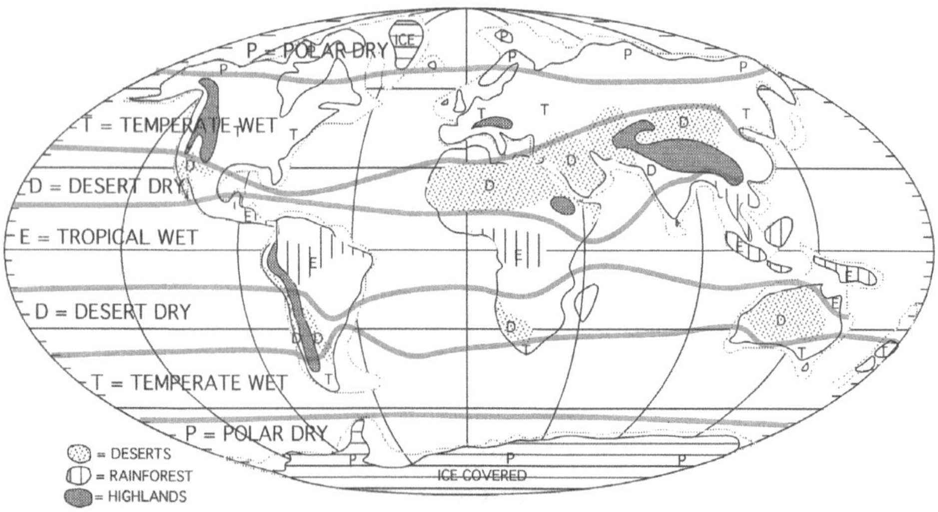

Figure 2 Generalized modern climate zones.

Cross-latitudinal ocean currents: The poleward flow of warm-water ocean currents along the **eastern margin** of continents (Figure 3) produces **climate warming** in the areas affected, while the equatorward flow of cold-water ocean currents along the **western margin** continents produces **climate cooling**. The impact of this effect varies, but all North Atlantic areas bathed by Gulf Stream waters have warmer climates than indicated by their latitudinal position.

Cold plateaus: High plateaus have **cold climates** because of the altitudinal temperature gradient. High-altitude air is less dense, therefore cooler than low-altitude air. High plateaus create their own climates and exert influence over large regions. They also tend to have drier climates.

Deserts: Deserts form where there is an excess of evaporation over rainfall. There are two major locations for deserts: low-latitude position and remote interiors of continents (Figure 2). Latitudinal deserts are predictable from paleolatitude reconstructions of atmospheric circulation cells, while interior deserts are predictable from continental reconstructions.

Latitudinal deserts: There are two zones of hot deserts (which includes most of the large deserts) corresponding with high-pressure ridges from 15° to 35° N and S latitude (with variation). There are cold areas of desert-like low precipitation at the north and south poles (80°–90° N and S) where descending air masses form high-pressure centers.

Continental interior deserts: Interior deserts occur in the interiors of large continents because they are distant from oceans, which are the source of moist air that produces rain. Air that moves great distances over land becomes depleted in water vapor. The large deserts of central Asia are continental interior deserts and the large high-standing supercontinents of the Mesozoic also had continental interior deserts.

Rain shadow deserts: Rain shadow deserts are formed by similar processes.

Sediment and Fossil Paleoclimate Indicators

There are several types of sediments that are reliable indicators of climate. These are the products of limited distribution of sediment-producing organisms or temperature and water-dependent processes such as the accumulation of glacial ice or formation of soil type.

Reefs—Indicators of hot and warm climates. The distribution of reef-forming organisms is temperature controlled. Reef corals occur in waters ranging from 20° to 34°C and are unable to survive seasonal temperature shifts below 17°C, so the reef community is a good indicator of tropical marine environments, in areas with low rates of sediment input. Organisms that formed ancient reefs also occurred at low paleolatitudes, showing that

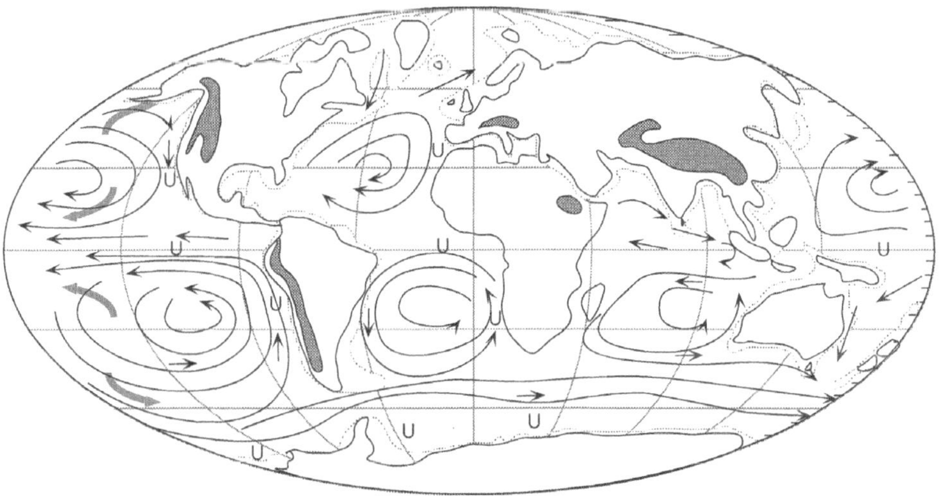

Figure 3 Circulation pattern of ocean surface waters, with directions of the driving winds shown on the left margin.

the occurrence of reefs can be used as an indicator of warm marine climates. **Carbonate sediments** are sometimes used as indicators of warm climates, but carbonate-producing organisms, and therefore carbonate sediment distribution, occur over a wider range of latitude than reefs. Some carbonate sediment deposition occurs in temperate climates.

Evaporites—Indicator of hot-dry climates. Sedimentary deposits of evaporite minerals are excellent indicators of hot-very dry (arid) climates. Evaporite minerals precipitate from high-salinity waters in hot desert climates, where rates of evaporation are high. These correspond with the low latitudinal high-pressure ridges and interior desert regions. **Red beds**, sediments composed of mineral grains coated with a layer of red-colored iron compounds such as hematite, are sometimes used as indicators of warm climates. Red beds are most common in warm dry climates, where porous sediments remain within the zone of oxidized ground waters for a long time. However, red beds are diagenetic in origin, so they may form under a different climatic regime than that present at the time of deposition. Therefore, red beds are less reliable as paleoclimate indicators.

Coal—Indicator of wet climates. Coal (and its precursor, peat) accumulates in areas where the decomposition plant remains is inhibited by the presence of standing water (as in swamps) or a very high water table. Conditions for peat accumulation correlate closely with areas of high rainfall in low-pressure troughs, especially in the hot-wet climates zone near the equator (where most modern peat accumulation occurs) and the cool-wet climates of high temperate latitudes (such as the northern peat bogs of Maine, Ireland, and Russia). On a cool Earth having climates like those of today coal accumulation occurs preferentially near the equator. On a hot Earth, coal accumulation tends to occur away from the equator.

Tillites—Indicators of cold climates. Till is an ice-margin deposit of glaciers, so the occurrence of tillite is a good indicator of cold climates. Till is a reliable indicator of polar and continental glaciers, because the till of alpine glaciers (the other source of till) has little chance of being preserved. Till has very poor sorting and often contains striated grains and cobbles. Another indicator of glaciers is striated pavement, which provides evidence of the direction of ice movement.

Fossil soils (paleosols). The type of soil produced in an area is strongly controlled by the amount of rainfall in the area, so different types of paleosol can distinguish between areas of high rainfall, low rainfall, and seasonally variable rainfall (monsoonal climate). Soils of high-rainfall areas are leached of soluble materials while soils of low-rainfall areas have high concentrations of carbonate minerals. Soils formed in areas of monsoonal climates usually have shrinkage-expansion fabrics that are diagnostic, and may have carbonate concentrations. Laterite soils (which are highly leached of solubles) are characteristic of tropical hot-wet climates.

Exercise Work

Describing paleoclimates

Reconstructing global paleoclimates utilizes predictive modeling and generates testable predictions in a series of iterative steps. The first step uses primary controls of temperature, wind, and rainfall on continents and ocean basins and then secondary controls are added until a reasonable reconstruction is obtained, that is checked against paleoclimate indicators to produce a finished product. Use modern patterns of climate zones as a reference to reconstruct ancient climates.

1. On a paleogeographic map of continents and ocean basins for each geologic period, **draw** in the general pattern of **surface ocean currents (gyres)** expected for the paleogeographic arrangement of continents. The latitudinal limits of prevailing surface winds are marked on the left side of the map. **Start at the equator** and reconstruct ocean currents for both hemispheres. In general, symmetric surface circulation cells (gyres) form on either side of the equator in large ocean basins. Use the patterns of surface winds (the

driving force) to determine the movement of ocean current gyres. Where continents interrupt latitudinal flow of ocean currents, indicate the directions in which surface waters are deflected to form large cross-latitudinal currents. Ocean currents in gyres move warm waters poleward from tropical regions on the west sides of ocean basins (like the Gulf Stream) and move cold waters equatorward on the east sides of ocean basins (like the Peruvian current).

2. On the same paleogeographic map, lightly **draw** in the predicted boundaries of the latitudinal climatic zones for hot and wet, hot and dry, cool and wet, and cold and arid climates.

   A. Draw in boundaries for the hot and wet climate zone present on or near the equator and label the area as **hot–wet.**

   B. Draw in boundaries for the hot and dry climate zones present in the Tropic paleolatitudes and label the areas as **hot–dry**. In areas with no highlands or strong cross-latitudinal ocean currents, mark the hot-wet and hot-dry at 25° North and South latitude to fit the latitudinal width of the Hadley circulation cells.

   C. Draw in boundaries for the cool and wet climate zones of mid-high paleolatitudes and label the areas as **cool–wet**. Adjust boundaries to accommodate deserts formed in continental interiors and downwind of high mountain ranges.

   D. Draw in the boundaries for cold and arid climate zones at the poles and label the areas as **cold–arid**.

3. **Determine** where ocean currents could modify these climate zone boundaries along the eastern and western margins of continents and **adjust** the boundaries for the provinces.

4. Add the following to the maps:

   A. **Determine** where continental interior deserts might be located. Label these climate areas as **ID** (interior deserts) and **draw** boundaries for the province. Continental interior deserts form in the center of large continents, in areas distant from oceans, because prevailing winds are depleted in moisture as they move long distances over land.

   B. On the same paleogeographic map, evaluate the placement of highland plateaus and mountain ranges (shaded areas on maps) and **determine** where cold plateau climates might be located. Label these climate areas as **CP** (cold plateau) and **draw** boundaries for the province.

5. On the same paleogeographic map, **show** your predictions of where east coast rainy areas (**ER**), west coast deserts (**WD**), continental glaciation (**G**), or reefs (**R**) might occur, using the criteria discussed below.

   A. East coast rainy areas (**ER**) develop because low-latitude trade winds bring ocean moisture over the eastern sides of continents, producing large areas of wet climate on the eastern sides of low-latitude continents.

   B. West coast deserts (**WD**) develop because low-latitude trade winds pull air from eastern sources over the continents. Rainfall over land causes the air to become progressively depleted in moisture, producing an east–west rainfall gradient.

   C. Continental glaciers (**G**) form where climates are cold and precipitation exceeds summer snow melt. Levels of precipitation in polar regions can be estimated by determining whether surface winds travel over open ocean (high precipitation) or ice fields (low precipitation).

6. The second map of each set contains data used to determine climate. Compare maps of predicted climate zones with maps of the same time period showing paleoclimatic data. See if the patterns of data match your predictions of climate zones. Describe any major differences in the climate zones you predicted (first

maps) and the patterns you produced on the maps using paleoclimate indicators and paleogeography and circulation systems. Where are the differences and what are they?

Carboniferous

Cretaceous

7. A source of quantitative data for determining paleotemperatures comes from oxygen isotope ratios of minerals precipitated in the environment. Using the chart shown in Figure 4, convert O_{18}‰ data to paleotemperatures and determine a mean annual temperature for each climate zone.

   Carboniferous: 80°S: +3‰; 60°S: +1‰; 30°S: −2‰; equator: −3.2‰; 30°N: −2.1‰; 60°N: −0.5‰; 80°N: +2.5‰

   Cretaceous: 80°S: +1.2‰; 60°S: −1.5‰; 30°S: −2.5‰; equator: −4‰; 30°N: −3.2‰; 60°N: −1.8‰; 80°N: +1‰

8. Using the final maps of paleoclimates that you have constructed describe the climatic condition of **high paleolatitude** and **polar regions** for the Earth during the:

   Carboniferous:

   Cretaceous:

9. When was there continental glaciation and where was it located?

10. When were there large deserts present and where were they located?

11. When was there extensive deposition of coals?

12. During Cretaceous hothouse climates, was there much deposition of coal in the equatorial zone?

13. During each time period examined, did the Earth have **hothouse** or **icehouse** global climates?

    Carboniferous:

    Cretaceous:

14. Assuming that equatorial temperatures are similar through the Carboniferous to Recent interval of time, when were global temperature gradients highest (greatest temperature difference between equator and poles) and when were they lowest?

    Highest:

    Lowest:

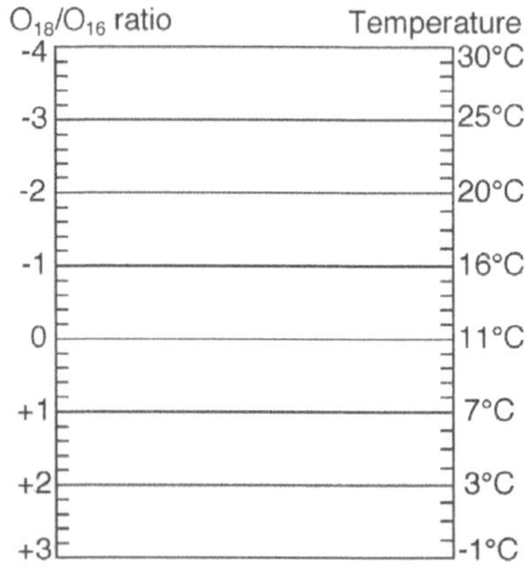

Figure 4 O_{18}‰-°C temperature conversion chart.

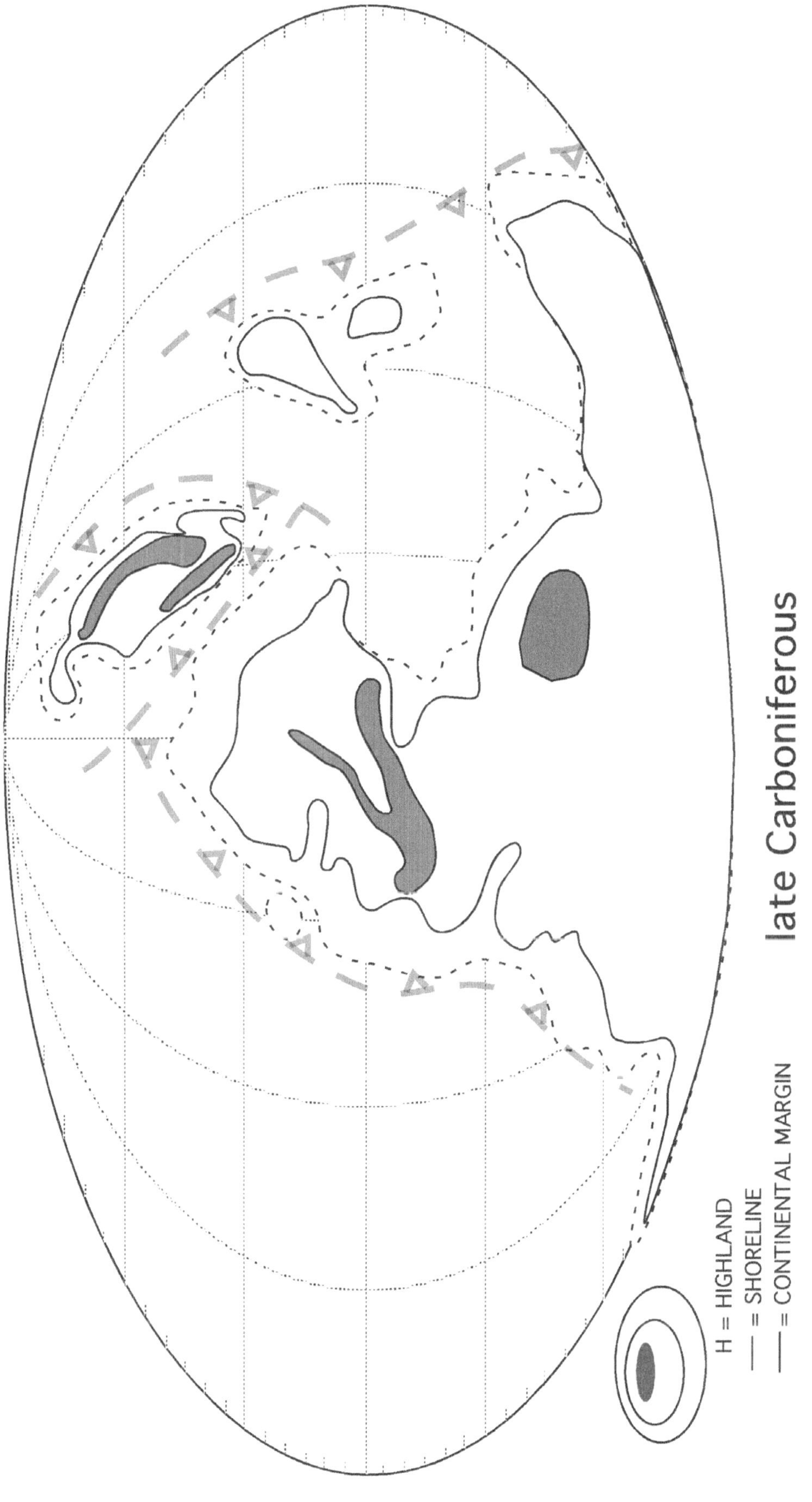
late Carboniferous
H = HIGHLAND
= SHORELINE
= CONTINENTAL MARGIN

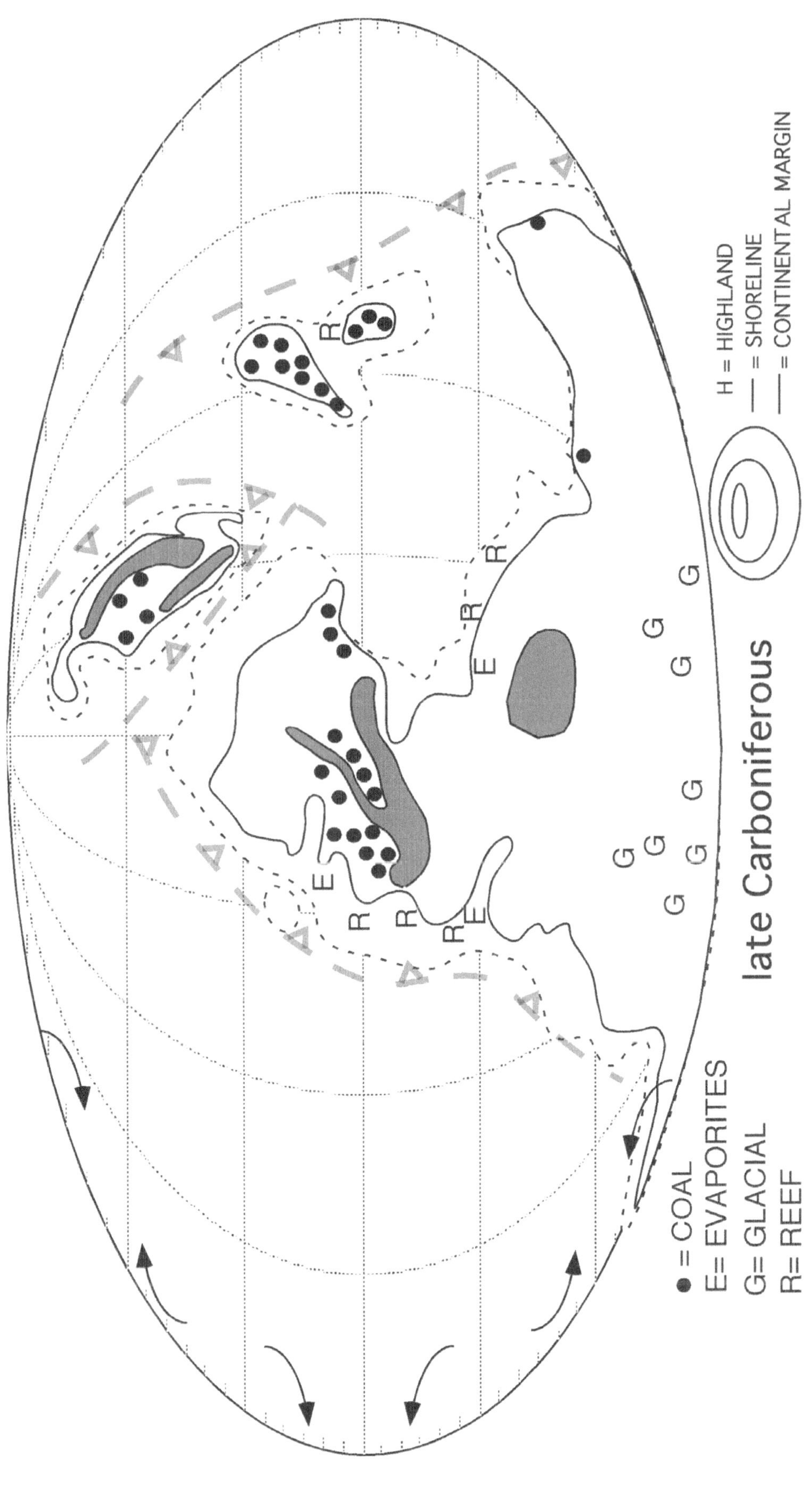
H = HIGHLAND
= SHORELINE
= CONTINENTAL MARGIN
late Carboniferous
● = COAL
E= EVAPORITES
G= GLACIAL
R= REEF
E
R
G

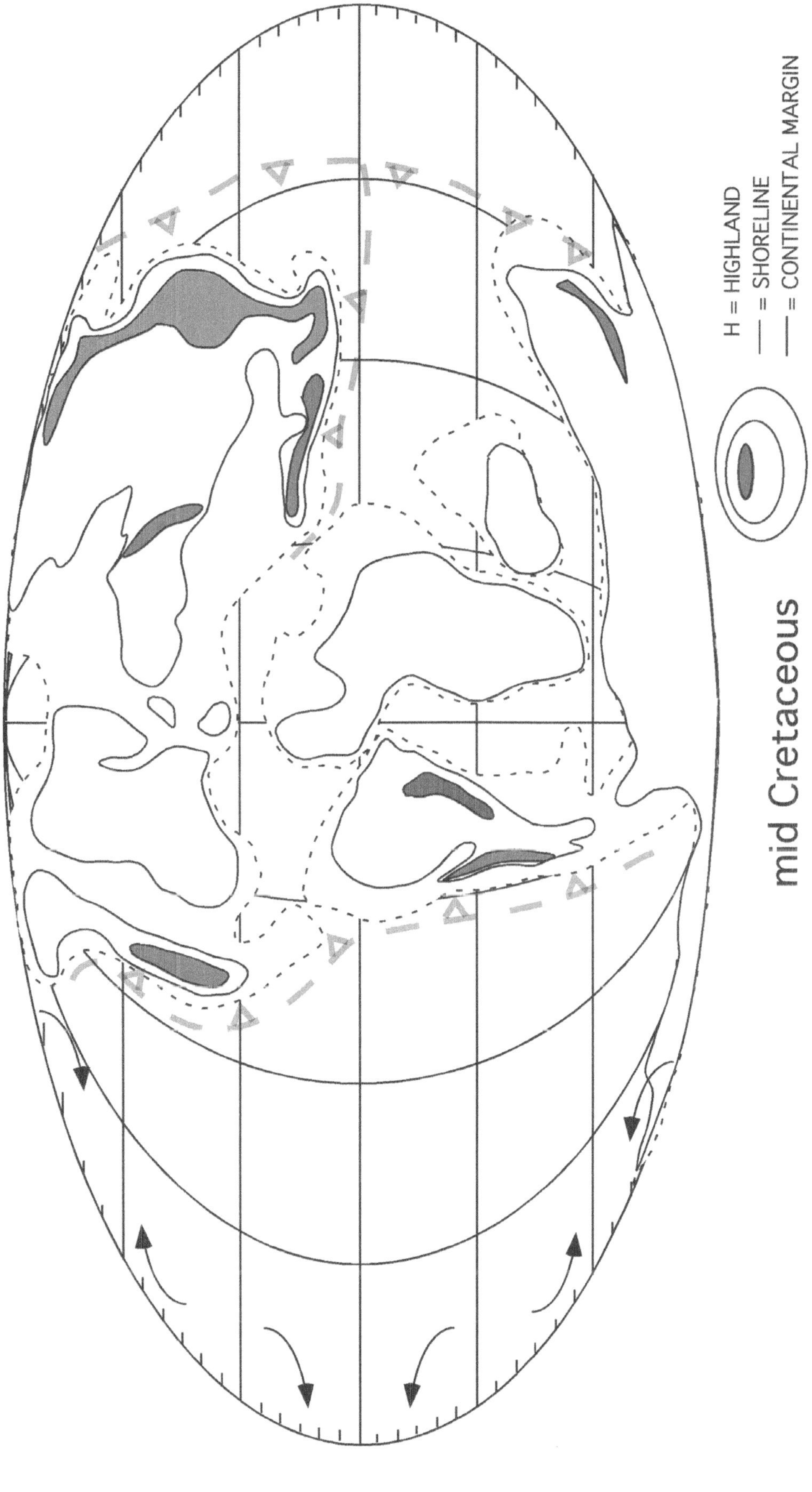
H = HIGHLAND
= SHORELINE
= CONTINENTAL MARGIN
mid Cretaceous

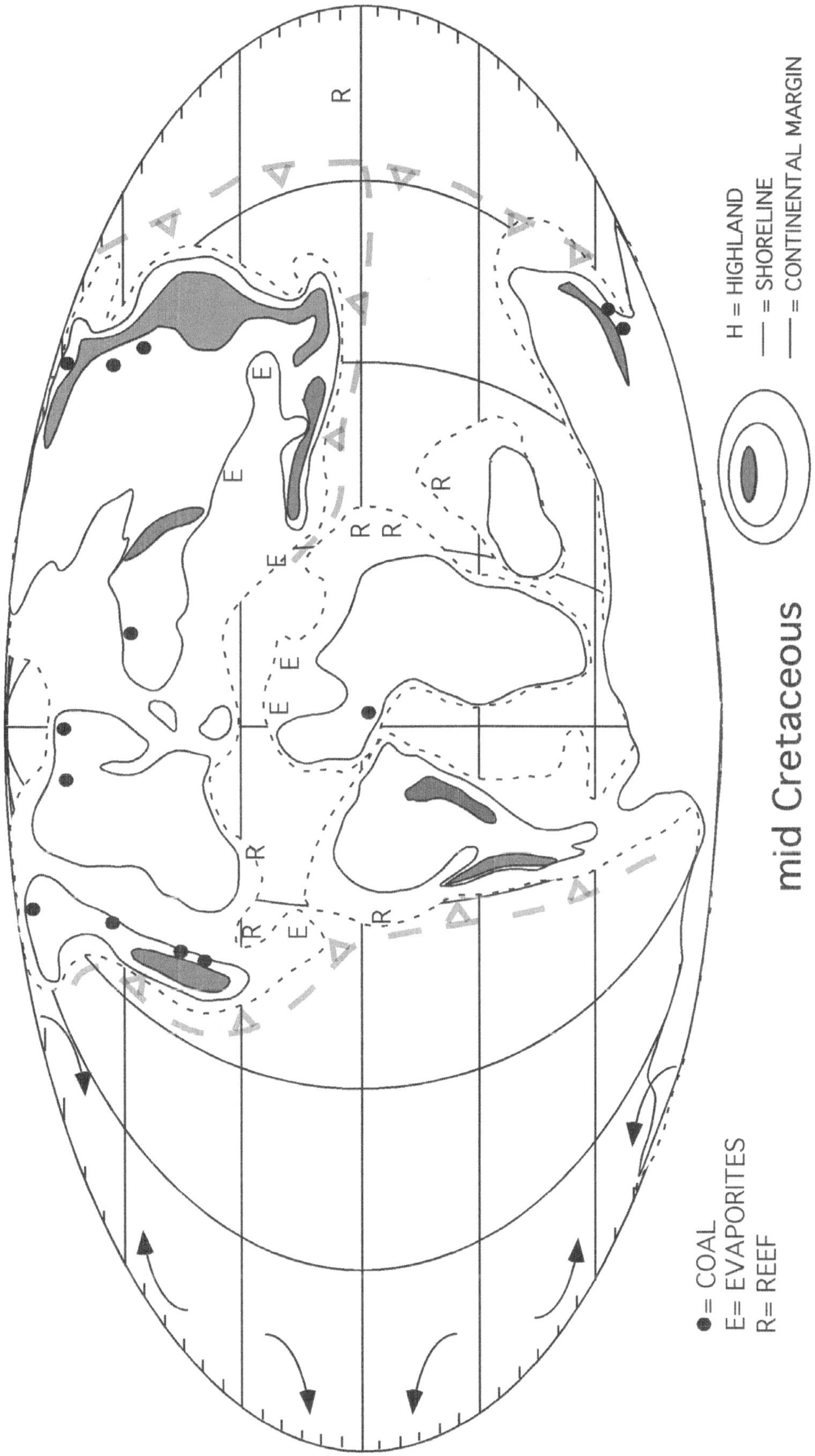
mid Cretaceous
● = COAL
E= EVAPORITES
R= REEF
H = HIGHLAND
= SHORELINE
= CONTINENTAL MARGIN